IB BIOLOGY MODEL ANSWERS

This model answer booklet is a companion publication to provide answers for the activities in the IB Biology Student Worktext (3rd edition). These answers have been produced as a separate publication to keep the cost of the workbook itself to a minimum. Working and explanatory notes have been provided where clarification of the answer is appropriate.

PHOTOCOPYING PROHIBITED

including photocopying under a photocopy licence scheme such as CAL

Third edition 2024

ISBN 978-1-99-101423-8

Copyright ©2024 Richard Allan
First published 2024
by BIOZONE International Ltd

First printing
Printed by Thomson Press

Acknowledgements

BIOZONE wishes to thank and acknowledge the team for their efforts and contributions to the production of this title.

Cover Photograph

Photo: iStock - Adam Bennie

The snowy owl (*Bubo scandiacus*) is a migratory bird inhabiting the Arctic tundra in warmer months and migrating south to North America, Europe, and Asia in winter. It is diurnal, hunting its primary food source (lemmings) day and night. Snowy owls breed on the Arctic tundra and are highly territorial, defending the nest vigorously against much larger animals, including wolves. Its magnificent plumage ranges from snowy white in older males, to white with dark bars and spots in females and juveniles.

BIOZONE International Ltd.
32 Somerset Street, Frankton,
Hamilton 3204, New Zealand

PH: +64 7 856 8104
FAX: +64 7 856 9243
Email: sales@biozone.com

www.BIOZONE.com

Contents

Theme A: Unity and Diversity

Chapter 1: Molecules

Chapter 2: Cells

Chapter 3: Organisms

Chapter 4: Ecosystems

Theme B: Form and Function

Chapter 5: Molecules

Chapter 6: Cells

Chapter 7: Organisms

Chapter 8: Ecosystems

Theme C: Interaction and Interdependence

Chapter 9: Molecules

Chapter 10: Cells

Chapter 11: Organisms

Chapter 12: Ecosystems

Theme D: Continuity and Change

Chapter 13: Molecules

Chapter 14: Cells

Chapter 15: Organisms

Chapter 16: Ecosystems

Theme A: Unity and Diversity

Chapter 1: Molecules

1. Water in Living Systems (Page 3)

1 a. *Water has hydrogen bonds between each molecule that hold the water molecules together. These also account for its high heat capacity as they store a lot of energy.*

b. *Water's polarity means it can attract both positive and negative ions and molecules. This means it can break up solutes into individual molecules or ions and hold them in solution.*

c. *Large amounts of energy must be put in to break the hydrogen bonds between water molecules. It takes a lot of energy to change water from liquid to gaseous state.*

2. *When water freezes, it forms a lattice with large spaces between the molecules. In liquid form, the molecules pack together more closely, giving liquid water a higher density.*

3. *Water is a liquid at normal room temperature and pressure and acts as a solvent. This allows it to support aquatic life and provide an internal environment in which biochemical reactions can occur. The amount of energy it can absorb before changing state buffers bodies and ecosystems against extremes of temperature. It is colourless so allows photosynthesis to occur in aquatic systems.*

4. *Marine mammals normally have thick insulation, either fur or blubber to reduce heat loss while in the water. If out on land, this insulation can actually cause overheating as heat is lost to the air more slowly than to water.*

2. The Origin of Earth's Water (Page 4)

1. *Water is a solvent for many molecules and is important as a medium for the chemical reactions of life.*

2. *Liquid water is maintained on Earth's surface because of Earth's distance from the Sun (water neither boils away nor is permanently frozen). Earth has sufficient mass to retain its water and atmospheric pressure to keep water on its surface.*

3. *The hydrogen/deuterium ratio of water found on asteroids is similar to that found in the oceans on Earth. The hydrogen/deuterium ratio found in water in comets is higher than found in Earth's oceans so it is unlikely they provided much water.*

4. *It is the zone around the Sun in which conditions for life are just right, i.e. the temperature is warm enough to stop water freezing but cool enough to stop it from boiling away into space.*

3. Nucleotides and Nucleic Acids (Page 6)

1 a. *Base pairing always follows the rules: C with G, G with C, T with A, and A with T.*

b. *In mRNA uracil replaces thymine (U pairs with A)*

c. *The hydrogen bonds in double stranded DNA hold the two DNA strands together.*

2. *Nucleotides are the building blocks of DNA. Their precise sequence provides the genetic blueprint for the organism.*

3.

	DNA	RNA
Sugar present	*Deoxyribose*	*Ribose*
Bases present	*A, T, C, G*	*A, U, C, G*
Number of strands	*2 (double strand)*	*1 (single strand)*

4. *The bases can be in any order and a DNA sequence can be thousands or millions of bases long. This means information can be stored in the order (sequence) of bases and the number of bases in DNA.*

5. *DNA is found in every living organism. Its makeup and structure are the same in all life. We can conclude that this must be because it was present in the very first organisms and was inherited from them.*

4. DNA and Directionality (Page 7)

1 a. *The asymmetric bonds in the sugar-phosphate backbone give the molecule a direction so that the strands in the double-helix run in opposite directions.*

b. *5' end terminates in a phosphate group, 3' end terminates in a hydroxyl group (from a sugar).*

2. *Hydrogen bonding helps, at least partially, with the stability of the DNA molecule. AT pairs are bonded by 2 hydrogen bonds, while CG pairs are bonded by 3 hydrogen bonds. DNA of organisms living in high temperature environments tend to have a greater amount of CG pairs in their DNA. Areas of DNA that need to be separated frequently tend to have more AT pairs.*

3. *The length of a CG base pair is equal to the length of an AT base pair. Thus, width of the DNA molecule is maintained, no matter the number of order of these base pairings.*

5. Creating a DNA Molecule (Page 8)

2.

Template strand	Complementary strand
Cytosine (C)	Guanine (G)
Guanine (G)	*(a) Cytosine (C)*
Thymine (T)	*(b) Adenine (A)*
Adenine (A)	*(c) Thymine (T)*
Thymine (T)	*(d) Adenine (A)*
Adenine (A)	*(e) Thymine (T)*
Thymine (T)	*(f) Adenine (A)*
Thymine (T)	*(g) Adenine (A)*
Cytosine (C)	*(h) Guanine (G)*
Guanine (G)	*(i) Cytosine (C)*

ISBN: 978-1-99-101423-8

6. *The DNA strands run next to each other but in opposite directions (anti-parallel).*

7. *Bases (A,T,C,G) always pair following the rule: A pairs with T and C pairs with G. Using this rule, it is possible to construct the DNA since the bases on one strand are already known. It is simply a matter of pairing any A or T on the known strand with T and A on the incomplete strand. The same with C and G.*

6. The DNA Molecule (Page 11)

1. *DNA is a very long molecule. If it was not packaged up it could not fit into the nucleus and it could not be properly organized for replication.*

2. *Histones provide a multi-unit structure for the DNA to wrap around in an organized way so that it takes up less space in the nucleus than if it was spread out.*

3. *Coiling the chromosome into an orderly, tight structure helps the cell maintain order and ensures proper segregation of the chromosomes during mitosis and cell division.*

4. *The packaging proteins (histones) can be modified in various ways so that DNA can be packed loosely or tightly. When tightly packed, the genes cannot be expressed.*

7. Investigating the DNA Molecule (Page 12)

1 a. *The radioactive protein did not enter the cell so it could not have caused the transformation. The transforming factor must have been the DNA as it entered the cell.*

b. *The assumption made was that DNA is the same for viruses, bacteria, and eukaryotes.*

c. *If protein was responsible for carrying genetic information, it would have entered the cell.*

2. *(Answers Only)*

Species	Ox		Human		Yeast	Bacilli bacteria
Tissue	Thymus	Spleen	Thymus	Liver		
Adenine						
Guanine						
Cytosine						
Thymine						
A/T		1.1	1.0	1.0	1.0	1.1
C/G		0.80	0.84	1.4	0.88	0.93
A+T/C+G		1.6	1.9	2.0	2.1	0.77

3. *The ratio is the same within a species (in different organs) but different between species.*

4. *1. In DNA, amount of A is equal to the amount of T and the amount of C is equal to the amount of G.*

   2. The amount of A + T does not equal the amount of C+G (the ratio is not 1:1).

5. *The tetranucleotide hypothesis predicted that the amount of A, T, C, and G should be the same because the bases repeat in groups of four. The data does not support this. The data also shows that, because the amounts of A, T, C, and G are different, DNA has the potential to store information as varying sequences of bases (the tetranucleotide hypothesis predicts DNA cannot store information as a varying sequence of bases).*

8. Did You Get It? (Page 14)

1. *Hydrogen bonds are relatively strong intermolecular bonds. They account for water's strong surface tension and its adhesive, solvent and thermal properties. In all these cases, the hydrogen bond is strong enough to hold water molecules together between 0°C and 100°C, but weak enough to allow water molecules to move around. Beyond these temperatures the hydrogen bond is strong enough to stop movement (freezing) or weak enough to allow molecules to separate (vaporization).*

2 a. *Adhesive property*

b. *Cohesive property*

c. *Thermal property*

d. *Solvent property*

3 a. *A (adenine)*

b. *G (guanine)*

c. *T G C T A G A G T G T T*

4.

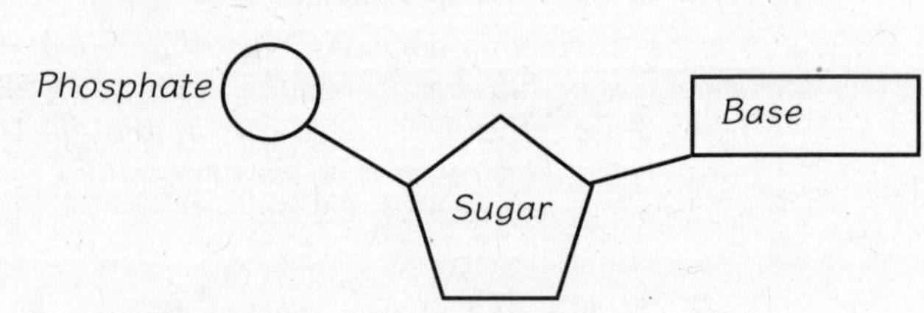

5. *Any two of: DNA is 2 strands RNA is 1, DNA bases are ATCG but RNA bases are AUCG, DNA molecules are long but RNA molecules are short, DNA has a deoxyribose sugar but RNA has a ribose sugar.*

6. *Water plays essential roles in metabolism as we know it. It dissolves molecules and ions and acts as a medium for chemical reactions to take place as well as taking part in, or being the result of, chemical reactions. Thus, it must be present for life as we know to exist.*

7 a. *Hydrogen bond*

b. *2*

c. *3*

8. *A = T so T = 29%*
C + G = 100% - (A + T) = 100 - (29 + 29) = 42
C = G so 42/2 = 21 so C = 21%, G = 21%

ISBN: 978-1-99-101423-8

Theme A: Unity and Diversity

Chapter 2: Cells

9. The Origin of Life On Earth (Page 17)

1. *Methane and carbon dioxide.*
2. *High temperatures and UV radiation could have stimulated formation of organic molecules.*
3. *RNA*
4. *The protobiont is regarded as a primitive cell because it has a double layer lipid membrane, similar to modern cells, which encloses self replicating RNA.*

10. Prebiotic Experiments (Page 18)

1. *Atmospheric conditions similar to those 3.8 billion years ago would be able to produce the complex chemicals required for the origin of life.*
2. *Yes, the Miller-Urey experiments did support the theory that conditions on early Earth could produce simple organic molecules. When the Earth's conditions were replicated, amino acids, nucleic acids, sugars, lipids, and ATP were formed.*
3. *Present day Earth conditions and materials are unable to produce the complex chemicals needed. Current life would out-compete new life.*

11. An RNA World (Page 19)

1. *Ribozymes can act both as genes and enzymes. This allows us to construct a plausible model for the origin of life because the ribozyme RNA molecules could perform the catalytic activity needed to assemble themselves.*
2. *Mutations in one RNA template could provide advantages over another RNA template, e.g. by enabling faster replication or producing proteins that helped copying. The most 'efficient' RNA would become predominant.*

12. The Common Ancestry of Life (Page 20)

1. *All organisms use the same nucleotide bases in their DNA and these bases all code for the same amino acids.*
2. *From the commonality in the genetic code and from the similarities in the molecular machinery of all cells.*
3. *The types of energy sources found around deep sea hydrothermal vents are likely to be similar to the energy sources used by early life forms when oxygen was not available.*

4 a. *Stromatolites are rock formations formed of stratified sediments composed largely of cyanobacteria.*

b. *Stromatolites are ancient formations that have been present for billions of years. They contain fossilized remains of early life forms, allowing us to research what ancient life may have been composed of.*

5. *Student's own answer.*
Answer will vary, but examples could include: New Zealand tuatara (Sphenodon punctatus), Ginko tree (Ginkgo biloba), or Coelacanth (Coelacanthus granulatus).

Student answer could simply state the organism and when it first appeared on Earth, or may include more detailed information such as distribution and change over time. Suggested answers for both approaches is provided below: New Zealand tuatara (Sphenodon punctatus). Ancestors with similar features to the tuatara were first recorded in the fossil record ~250-225 million years ago.

Longer answer
Ancestors with similar features to the tuatara (Sphenodon punctatus) were first recorded in the fossil record ~250-225 million years ago. The tuatara is the only living member of the reptilian order Rhynchocephalia (Sphenodontia). The order was once widespread across Gondawana, but all species within the order (except the tuatara) become extinct about 60 million years ago. The tuatara is only found in New Zealand. Introduced predators and human activity have reduced its numbers and habitat..

Fossil remains show the tuatara has changed very little in its morphology (appearance) over time.

13. The Cell is the Unit of Life (Page 22)

1. *Lacking in membrane-bound nucleus or organelles. Relatively small. Circular, naked chromosome.*
2. *The presence of a membrane-bound nucleus and organelles. Linear chromosomes. Relatively large.*
3. *They can only replicate by using the mechanisms in cells they infect.*

14. The Light Microscope(Page 23)

1 a. *Eyepiece lens*

b. *Arm*

c. *Coarse focus knob*

d. *Fine focus knob*

e. *Objective lens*

f. *Mechanical stage*

g. *Condenser*

h. *Built-in light*

2 a. *600 X*

b. *600 X*

3. *31 - 22 = 9 9 x 3.33 µm = 29.97 µm*
4. *43 mm = 43,000 µm*
Magnification = size of the image ÷ actual size of object = 43,000 µm ÷ 2 µm = 21,500 x magnification

15. Preparing a Slide (Page 25)

1. *Thin sections allow light to pass through because they have fewer layers of cells to interfere with viewing.*
2. *The onion epidermal cells do not take part in photosynthesis so they do not contain chloroplasts.*

ISBN: 978-1-99-101423-8

16. Developments in Microscopy (Page 26)

1. *Any of: it is portable, inexpensive, easy to learn to use, can be used to view live organisms.*
2. *The electron microscope allows for a much higher resolution of images. In cellular microscopy the images allow scientists to examine the cell at organelle level. The images are also digitized.*
3. *The ability for samples to be frozen in a vitreous (glass-like) state that removes image distortion.*
4. *Without the stain, features of the onion cell are difficult to see. The iodine stain produces greater contrast and makes the nucleus and the cell wall of the onion stand out and easier to see.*
5. *Stains are mostly used to enhance specific features of a sample, e.g. specific organelles.*
6. *Viable stains are harmless and can be used on living tissue. Non viable staining is used on cell or tissue preparations that are dead.*

7 a. *Trypan blue*
   b. *H&E*
   c. *Immunofluorescence*
   d. *Methylene Blue*

17. Common Features of Cells (Page 28)

1. *Cells need a source of carbon to build organic molecules. In plants, this is carbon dioxide. In animals, carbon comes from organic molecules including carbohydrates. Cells need ions to help maintain essential metabolic functions such as biochemical reactions and membrane transport. These ions may be needed as components of enzymes or transport systems such as ion pumps.*

18. Prokaryote and Eukaryote Cells (Page 29)

1. *Prokaryotes have no nucleus. The genetic material is a single, circular chromosome. Ribosomes are 70S; membrane bound organelles, e.g. mitochondria, endoplasmic reticulum are missing. Circular plasmids are usually present. Cell wall is chemically different from eukaryotic cell walls.*
2. *Have membrane bound organelles; membrane bound nucleus. Multiple linear chromosomes made of DNA. Highly organized cell structure, compared to prokaryotic cells.*
3.

Structure of a generalized prokaryotic cell

19. The Processes of Life in Unicellular Organisms (Page 30)

1 a. *The organism can detect and take action (have a cellular response) to changes in its internal or external environment.*
   b. *The organism can produce copies of Its DNA and itself or recombine genetic material with others of its species to reproduce new members of its species.*
   c. *The organism can process and remove metabolic wastes from itself.*
   d. *The sum of all the chemical processes in a cell that maintain its (living) self.*

2. *Approximately 3 times*
3. *The deeper layers of mud are so thick that oxygen cannot diffuse into them. They therefore tend to remain anaerobic and so microbes living there tend also to be anaerobic.*
4. *Viruses do not carry out all the core process of life. They have no metabolism of their own and only reproduce when they have infected a living organism.*
5. *The contractile vacuole expels water that enters the cell by osmosis and helps maintain the cell's (internal) solute concentration.*

20. Eukaryotic Cell Structures (Page 32)

1. *Vacuoles in plant cells are large and have important roles in turgor, storage, waste disposal, and growth. In animal cells, vacuoles are small with minor roles in endocytosis and exocytosis.*
2. *Made of different substances. Chitin in fungi (like insect exoskeltons) and cellulose in plants*
3. *Sieve tubes and erythrocytes: in sieve tubes, the cells are connected and are differentiated to form a continuous tube for sap movement (phloem) so the nucleus would take up space if present. In the erythrocytes, the cell requires maximum room to uptake oxygen into the haemoglobin, so the nuclei are also removed.*

ISBN: 978-1-99-101423-8

4. *Large, fused cells have multiple nuclei as a result of many cells joining together but retaining their nuclei. The nuclei control the area of the cell nearest to them, maintaining the functioning of the large cell.*
5. *Plant, animal, and fungal cells all have a membrane surrounding the nucleus containing their chromosomes in a separate area from the cytoplasm. They have complex organelles, also membrane covered, such as mitochondria (chloroplasts in plants), and endoplasmic reticulum etc. In prokaryotes, none of those structures would be present.*

21. Light Microscopy and Cells (Page 34)

1 a. *Prokaryotes have no distinct nucleus: a darkened patch or circle inside each cell. Also, no distinct organelles such as chloroplasts.*
b. *A darkened spot or circle that forms the nucleus in each cell / cellular unit. A 'brick-like' structure of cells if in plant tissue. Distinct, membrane-bound organelles such as chloroplasts.*
2. *The keratosis is on the top right and is arranged in a circular pattern. Keratin is a substance, not a cell, so there are no visible cellular structures, such as nuclei.*

22. Electron Microscopy and Cells (Page 35)

1 a. *Plasma membrane*
b. *Golgi apparatus*
c. *Secretory vesicle*
d. *Mitochondria*
e. *Smooth endoplasmic reticulum*
f. *Chromosomes*
g. *Ribosomes*
h. *Rough endoplasmic reticulum*
2. *Smaller organelles such as ribosomes, and the finer details of mitochondria and chloroplasts.*
3. *The extended plasma membrane - forming pseudopodia for amoeboid movement - the irregular shape of the cell would not allow it to fit easily with other cells to form a tissue.*
4. *Rough endoplasmic reticulum surrounds the nuclear membrane (interconnected with the nuclear membrane), and has ribosomes embedded in the membrane. Smooth endoplasmic reticulum does not have ribosomes embedded in the membrane; they function to synthesize lipids, phospholipids, and steroids.*
5. *Golgi apparatus processes proteins received from the endoplasmic reticulum into further products required by the cells. They organize them for transport and, wrapped in the phospholipid membrane, they break away from the end of the organelle. Once they have broken away, they become secretory vesicles that can move outside the cell.*
6 a. *Cytoplasm*
b. *Sap vacuole*
c. *Starch granule*
d. *Chloroplast*
e. *Mitochondria*
f. *Cell membrane*
g. *Chromosomes*
h. *Nuclear membrane*
i. *Endoplasmic reticulum*
j. *Cell wall*
7. *Sap vacuole - the products of photosynthesis, such as glucose, are dissolved in water and stored in these. They products can then be moved out of the cell for use elsewhere in the plant, used in the cell for cellular respiration, or converted to starch. Starch granules, a product formed from glucose and stored for long term use, can be converted back.*
8. *Student answer will depend on whether they have chosen to draw an animal or plant cell. Organelle functions have been provided as a table to keep the diagrams uncluttered.*

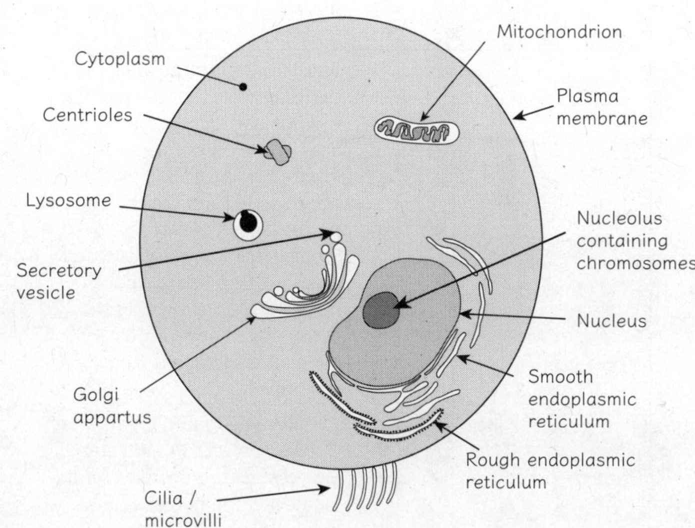

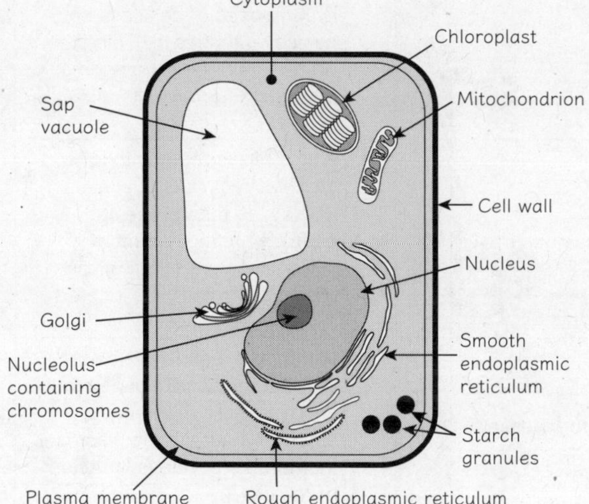

ISBN: 978-1-99-101423-8

Organelle	Plant	Animal	Function
Nucleus	✔	✔	A membrane structure containing the cell's chromosomes.
Mitochondrion (mitochondria plural)	✔	✔	The cell's energy producers. They use the chemical energy in glucose to make ATP through the process of cellular respiration.
Chloroplasts	✔		Found in plant cells, and autotroph protists. A specialized plastid containing the green pigment, chlorophyll. Chloroplasts are the site for photosynthesis.
Sap vacuoles	✔		Large vacuoles are for storage of sugar products from photosynthesis, dissolved in water.
Golgi apparatus	✔	✔	A structure made up of membranous sacs. It stores, modifies, and packages proteins for export from the cell.
Endoplasmic reticulum (ER)	✔	✔	A network of tubes and flattened sacs continuous with the nuclear membrane. There are two types of ER. Rough ER has ribosomes attached. These are the sites of protein synthesis in a cell. Smooth ER has no ribosomes attached and has a smooth appearance. Involved in the synthesis of non-protein molecules including steroid hormones.
Chromosomes	✔	✔	The genetic material of the cell. Consists of a DNA molecule and associated proteins, and located in the nucleus of eukaryotic cells.
Cell wall	✔		Not found in animal cells. A semi-rigid structure that lies outside the plasma membrane of plant cells (and fungi and some protists). In plant cells it is composed of cellulose and has several roles including: protecting the cell, providing strength and maintaining cell shape.
Plasma membrane	✔	✔	A lipid bilayer containing the cytoplasm of cells. It acts as a selective membrane and controls the movement of materials into and out of the cell. In plant cells it is located inside the cell wall.
Secretory vessels		✔	Membrane-bound vesicles that bud off the Golgi network and transport substances, e.g. hormones, to the plasma membrane for release outside of the cell.
Microvilli		✔	Small, finger-like extensions of the plasma membrane which increase the cell's surface area. Found in some animal cells.

23. Endosymbiosis Theory (Page 37)

1. *Prokaryotes have 70s ribosomes whereas eukaryotes have larger, 80S ribosomes in the cytoplasm of the cell so they must have undergone evolution to account for the change. Mitochondria and chloroplasts therefore resemble prokaryotes which provides evidence that they, along with their ribosomes, originated independently.*

24. Cellular Differentiation (Page 38)

1. *Zygote*
2. *All cells have the same genetic information as genes in the chromosomes. Cellular differentiation occurs due to differential gene expression: some genes are turned off (or on) to form the particular structures of different cells. Once this occurs, the final cell type is formed irreversibly.*

3 a. *Non-differentiated cells that can form any cell type.*

b i. *They can generate more stem cells (through mitosis).*

b ii. *They can differentiate into other cells.*

4. *Student answer: For example, sea turtles - incubation below 27.7 °C the hatchlings are all male, and above 31 °C the hatchlings are all female (note this is the opposite of American alligators). The higher temperature down-regulates (inhibits) the expression of the Sox9 gene that normally promotes male characteristics forming.*

25. Multicellularity (Page 40)

1. *The cells specialize, such as the heterocysts in the cyanobacteria that fix and share nitrogen with cells close by, and in the slime mould there are fruiting body cells and stalk cells. Each type of cell can only perform its own functions and not that of the other cells, but they all work together.*
2. *Student answer: For example, sponges - the sponges are a collection of specialized cells acting together, rather than made of distinct and organized tissue as in higher animals. The different cells, choanocytes, amoebocytes, porocytes, pinaocytes, each are specialized for a different function - and share resources with each other.*

26. Viruses (Page 41)

1. *Answer will vary depending on the virus students have chosen, but the report should include:*
   - Classification of the virus.
   - A scientific drawing of the virus, labelling key structures.
   - Identify its host(s).
   - Describe any impacts the virus has on the host.
   - Identify any impacts on humans (if any).

ISBN: 978-1-99-101423-8

27. Replication in Viruses (Page 42)

1. *A virus that can infect bacterial cells.*

2 a. *The virus glycoprotein spikes 'trick' the cell into recognizing it as non-harmful - the plasma membrane of the cell allows entry of the RNA. Viral nucleic acids are 'injected' into the cell, as well as enzymes - the other components of the virus remain outside.*

b. *As part of the bacterial genetic material - it combines and replicates at the same time the bacterial RNA/DNA replicates.*

c. *Also inside the bacteria - it uses the cellular mechanism (ribosomes etc.) to transcribe viral proteins and components that are assembled before release.*

3. *The process of creating new viral particles and assembling them and then breaking out of the cell to spread to other cells results in cell disruption and death. This cell death and the host cell's immune response to the pathogen are what cause disease.*

28. Rapid Virus Evolution (Page 43)

1 a. *The viral genome is contained on 8 short, loosely connected RNA segments. This enables ready exchange of genes between different viral strains and leads to alteration of the protein composition of the H and N glycoprotein spikes.*

b. *The body's immune system acquires antibodies to the H and N spikes (antigens) on the viral surface, but when different variants arise they are not recognized nor detected by the immune system (there is no immunological memory for the newly appearing antigens).*

2 a. *HIV has a very short generation time allowing, multiple chances for mutation.*

b. *RNA polymerase lacks error checking, thus allowing more mutations to occur.*

c. *An infected cell can produce large numbers of HIV every day, allowing large numbers of new viral variations to be produced.*

3. *140-304 generations (÷ 1.2 & 2.6 by 365)*

4. *Combination drug therapy targets many parts of HIV's replication pathway. By using a combination of drugs it slows down the replication and ensures that if the virus becomes resistant to one drug, a second and third will still work.*

29. Did You Get It? (Page 45)

1. *Eukaryote - the cell has a nucleus and presence of membrane covered organelles. Specifically animal - has no cell wall like fungi and plants. Has no chloroplasts like plant cells (or large vacuole).*

2. *The chromosomes would be free and contained in the cytoplasm (not in a nucleus) - and there would be a visible outside coat/wall around the cell - and no organelles visible.*

3. *These organelles have their own genetic material, as distinct from the cell's chromosomes in the nucleus, which replicates independently from that of the cell. The organelles have a double membrane (bilayer phopholipid) like the bacteria. The structures inside the organelles are similar (on a microscale) to bacteria.*

4. *The original observations were made on simple microscopes, and then later light microscopes, that could not use high resolution to enable clear viewing of the organelles and cellular structures inside the cell. It wasn't until a century later that electron microscopes were invented and allowed viewing at much greater resolution.*

5 a. *Genetic material of a virus.*

b. *The virus has only evolved to be detected by the tobacco plant - the human cells would not recognize the spikes (controlled by viral 'genes') and, therefore, it would not allow it access to the cell to cause damage / allow virus to replicate.*

6. *Lytic cycle - where the genes of the virus are being actively transcribed and translated into viral proteins, which are then assembled. The lysogentic cycle occurs when viral genetic material is replicated (along with the cells DNA/RNA) but not expressed (no protein made) - and represents a 'dormant' period.*

7. *The genetic material is replicated from one set in the parent cell into two separate identical sets, with nuclear membranes reforming around each set that moves to opposite ends of the cell. The cytoplasm is divided, with a new cell membrane forming between each (an a nucleus in each) - and separate. Mitosis is the process.*

8. *The cells can specialize into structures that perform a certain task that one cell alone could not do all by itself - some cells may die during the process (stalk cells), but overall the population will be advantaged.*

Theme A: Unity and Diversity

Chapter 3: Organisms

30. Variation in Organisms

1 a & b.

1	*Domain*	*Eukarya*
2	*Kingdom*	*Animalia*
3	*Phylum*	*Chordata*
4	*Class*	*Mammalia*
5	*Order*	*Primates*
6	*Family*	*Hominidae*
7	*Genus*	*Homo*
8	*Species*	*sapiens*

2 a. *Binomial nomenclature*

b. *Genus and species (generic and specific name).*

3 a. *Avoid confusion over the use of common names for organisms.*

b. *Provide a unique name for each type of organism.*

c. *Attempt to determine/define the evolutionary relationship of organisms (phylogeny).*

ISBN: 978-1-99-101423-8

4. <u>*Panthera tigris tigris*</u>

5 a. *The name is not italicized (<u>Ceratotherium simum</u>)*

b. *The species name is capitalized (<u>lupus</u>)*

31. What is a Species? (Page 49)

1 a. *A biological species is a group of organisms that can successfully interbreed to produce fertile offspring and are reproductively isolated from other such groups.*

b. *Some species of the <u>Canis</u> family can interbreed to produce fertile hybrids (thus contradicting the species definition).*

2. *Behavioural (they show no interest in each other).*

3. *Physical barrier; sea separating Australia from SE Asia.*

4. *The red wolf is rare and may have difficulty finding another member of its species to mate with.*

5. *They have very different songs; neither species recognizes the song of the other. They also have morphological/colouration differences.*

6. *As the greenish warblers spread east and west from the ancestral population, unique characteristics evolved in each subsequent population as a result of different selection pressures in each environment.*

7. *The greenish warbler shows species divisions to be arbitrary because although the two subspecies do not interbreed across the zone where they meet directly, the two populations are linked by gene flow around the ring to the south of this area. Thus, gene exchange does potentially take place between them, contrary to the biological species definition.*

32. Problems in Defining a Species (Page 51)

1. *Species concepts define species based on certain criteria. As such, when scenarios occur that are outside these criteria the concepts tend to fail. The biological species concept defines a species based on its members' ability to interbreed and produce viable offspring. However, this cannot be applied to extinct species, species that reproduce asexually or species that do form viable hybrids (i.e. what then is a 'species'?). Concepts like the phylogenetic species concept solve this problem, but end up with inconveniently larger numbers of closely related species.*

33. Karyotypes (Page 52)

1 a. *The layout of the karyotype, the full chromosome complement of a cell or organism, characterized by the number, size, shape, and centromere position of the chromosomes.*

b. *It can provide information on sex and chromosomal abnormalities*

2&3.

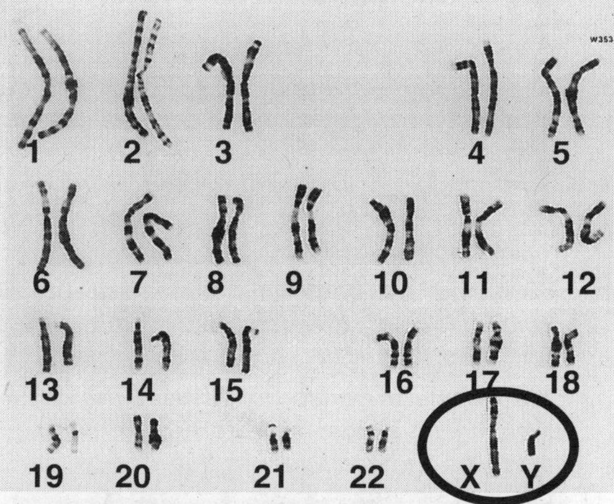

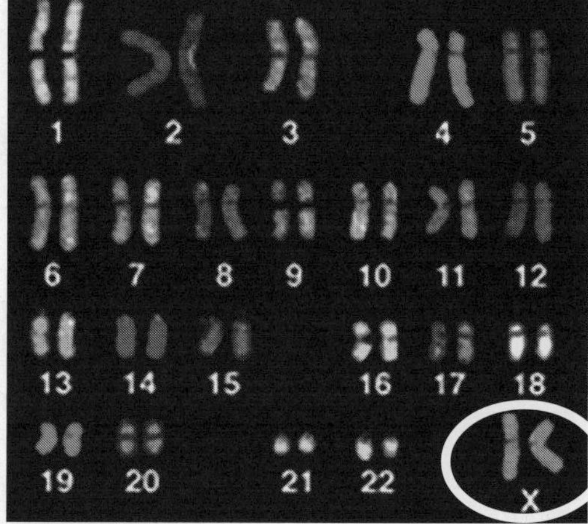

4 a. *44 XX*

b. *44 XY*

5 a. *46*

b. *23*

6. *Features include shape, size, centromere position, length of the arms, and banding pattern.*

7. *They are different in size, shape, banding pattern (and genes).*

8 a. *38*

b. *24*

c. *34*

34. Evolution of the Human Karyotype (Page 54)

1. *Scenario 2*

2. *Telomeres that are normally on the ends of the chromosomes are seen as remnants in the middle of human chromosome 2. The banding patterns of human chromosome 2 match the separate ape chromosomes. The human chromosome 2 has an inactive centromere (2 centromeres rather than 1).*

3. *If chromosome 2 split (instead of being produced by fusion), we would expect that human chromosome 2 would not have remnant telomeres roughly in its centre or remnants of a second centromere. We would also expect that any telomeres on ape chromosomes would likely be shorter on one end than the other on both new chromosomes. One of the new ape chromosomes would likely have a less developed centromere than the other.*

ISBN: 978-1-99-101423-8

35. Making a Karyogram (Page 55)

2. *Male*
3. *Diploid*

36. Diversity in Genomes (Page 58)

1 a. *All the genetic material in one haploid set of chromosomes.*
b. *A substitution of a single nucleotide at a specific location on the DNA.*
c. *A version of a gene - causes the production of a protein that differs slightly from allele to allele.*
2. *Variation within a species can be determined by studying the variation in SNPs in the genome. Studying the frequency of different SNPs or groups of SNPs (haplotypes) will provide information on relationships between individuals or can separate individual groups based on their specific SNP variations (haplotypes). The relationship between these haplotypes or their response to various scenarios can then be explored.*

37. Genome Size (Page 59)

1 a. *149,185 Mbp*
b. *490 Mbp*
c. *11.5 Mbp*
d. *0.12 - 0.21 pg*
e. *132.83 pg*
f. *2.81 pg*

38. Chromosomes and Species (Page 60)

1 a. *2*
b. *Horse roundworm*
c. *Potato*
2. *There is no correlation between the number of chromosomes an organism has and its complexity. Relatively simple organisms, e.g. the adder's tongue fern, can have a very large number of chromosomes, whereas very complex organisms, e.g. gorilla, can have far fewer.*
3. *The number of chromosomes, the banding patterns, and shape of the chromosomes are unique to each species and so can be used to define a species (different species' chromosomes will not match correctly).*

39. Using Whole Genome Sequencing (Page 61)

1. *The speed of genome sequencing has increased over time, especially after 2002 (almost exponential increase). The cost has decreased, especially after 2004.*
2. *Genome sequencing can be used for identifying and treating disease (e.g. identifying cancer prone genes), identifying the functionality of genes, identifying evolutionary relationships between species, and comparing genes between species.*

40. Classification Keys (Page 62)

1. *The case - presence or absence and specific features of the case itself.*

2 a. *Oxyethira*
b. *Hudsonema*
c. *Olinga*
d. *Aoteapsyche*
e. *Hydrobiosis*
f. *Helicopsyche*
g. *Triplectides*

3 a. *Plecoptera, stoneflies*
b. *Hemiptera, bugs*
c. *Coleoptera, beetles*
d. *Odonata, dragonflies/damselflies*
e. *Lepidoptera, moths & butterflies*
f. *Trichoptera, caddisflies*
g. *Emphemoptera, mayflies*
h. *Megaloptera, dobsonflies*
i. *Diptera, true flies*

41. Making a Classification Key (Page 64)

Student Investigation

42. DNA Barcodes (Page 65)

1. *DNA barcoding is a method of species identification by the assigning or identifying of unique, short DNA sequences, based on highly conserved sections, of DNA to each species.*

2 a. *Samples taken from the environment can be tested for specific barcodes from mobile, rare, or shy species.*
b. *Samples can be taken from areas such as waste water to test for pathogen barcodes that indicate the presence of a pathogen in a population.*
c. *Samples taken from the environment, e.g. water or droppings, can be tested for new or known barcodes to identify species in the environment.*

43. Classifying Organisms (Page 66)

1. *Molecular data has helped to separate groups that were once grouped together on the basis of appearance, e.g. DNA evidence supports splitting prokaryotes into the two domains, Bacteria and Archaea. The differences between these are as great as the differences between the prokaryotes and eukaryotes.*
2. *Classification helps us organize and recognize biodiversity.*
3. *Classification systems change as knowledge changes and new ideas advance understanding of relationships. The classification system changes to reflect these.*

4 a. *Advantages: They provide simple taxonomic ranks for organisms. Each organism is given a defining genus and species. Disadvantages: Does not always show correct evolutionary relationships (groups can be polyphyletic). Some ranks are not equivalent across groups, e.g. plants use divisions rather than phylums.*

ISBN: 978-1-99-101423-8

b. *Advantages: Shows the correct evolutionary relationships. All members of a group can be traced to a common ancestor. Takes all shared derived features both physical and molecular into accounts. Disadvantages: Groups may not have traditional taxonomic ranks (no specific naming system).*

5. *While traditional Linnaean taxonomy places organisms into groups based on their features and has a defined ranking system (kingdoms phylums etc.), cladistics has no set ranking system other than defining specific clades. Sometimes these clades will match the Linnaean system, e.g. mammals are a clade, whereas other times they will not, e.g. reptiles are not a clade, and clades produced from these organisms produce different groups from traditional classification.*

6 a. *A group descended from a single common ancestor with no other members.*

b. *A group of organisms descended from more than one common ancestor*

c. *A group descended from a single common ancestor but does not include all the descendants of that ancestor.*

7 a. *They are classified as reptiles because they all shared the features of scaly skin, cold-bloodedness (endothermic), laying eggs, etc.*

b. *This group does not include birds, which share an evolutionary ancestor with crocodiles (archosaurs). Also, lizards and snakes are as closely related to each other as crocodiles are to birds. Thus, reptiles is not a true clade .*

44. Cladograms and Phylogenetic Trees (Page 69)

1 a. *It is assumed that the simplest explanation is the correct explanation. E.g. a feature will only have evolved once and be passed on, rather than have evolved independently in many groups.*

b. *This would happen when a feature has evolved independently in different groups*

2. *A shared characteristic is one that is shared by two or more different groups. A shared derived characteristic is one that separates one group in the clade from the ancestor, setting it apart from other groups. E.g. birds all have feathers, derived from scales. This is a defining feature of birds.*

3 a. *The root is the point where the clade begins, the last common ancestor of all the members of the clade.*

b. *A point representing a common ancestor to the lineages that branch from it.*

c. *These represent the taxons, the groups defined by their relationships in the clade.*

4. *Match 3*

5. *The fact that figworts were grouped together based on their lack of characteristic features rather than features they shared.*

6. *The evolutionary history can be tested by examining shared derived features and genetic data. In this case, study of three chloroplast genes falsified the then interpretations of the figwort groups and showed that they were polyphyletic.*

7 a. *Phylogenetic tree 1.*

b. *It has fewer evolutionary events occurring than tree 2, so under the rule of parsimony we should accept this tree.*

c. *Event 2 occurred twice.*

45. Molecular Evidence and Cladistics (Page 72)

1. *Molecular clocks can help establish when species separated from each other. If a gene mutates at a certain rate, then counting the number of different mutations between species can indicate how long ago the separation occurred. Also, where the mutations occurred can help understand how the species are related.*

2.

SINE present (1)/absent (0)										
Taxon	**A**	**B**	**C**	**D**	**E**	**F**	**G**	**H**	**I**	**J**
Striped dolphin	1	1	1	1	1	1	1	0	0	0
Risso's dolphin	1	1	1	1	1	1	1	0	0	0
Indo-Pacific bottlenose dolphin	1	1	1	1	1	1	1	0	0	0
Common bottlenose dolphin	1	1	1	1	1	1	1	0	0	0
Long-beaked common dolphin	1	1	1	1	1	1	1	0	0	0
Chinese white dolphin	1	1	1	1	1	1	1	0	0	0
Pantropical spotted dolphin	1	1	1	1	1	1	1	0	0	0
Beluga	1	1	1	1	1	1	0	0	0	0
Finless porpoise	1	1	1	1	1	1	0	0	0	0
Yangtze River dolphin	1	1	1	1	1	0	0	0	0	0
Ginkgo-toothed beaked whale	1	1	1	1	0	0	0	0	0	0
Ganges River dolphin	1	1	1	0	0	0	0	0	1	0
Pygmy sperm whale	1	1	0	0	0	0	0	0	0	0
Omura's whale	1	0	0	0	0	0	0	1	0	0
Common minke whale	1	0	0	0	0	0	0	1	0	0
Hippopotamus	0	0	0	0	0	0	0	0	0	0

ISBN: 978-1-99-101423-8

3.

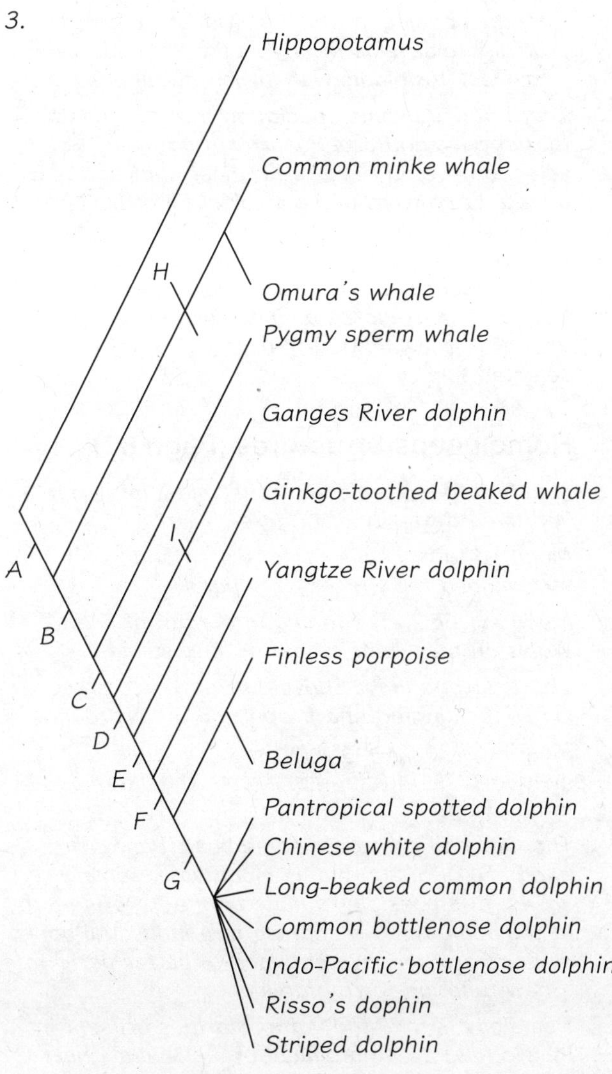

46. Did You Get it? (Page 74)

1. *A group of organisms that can interbreed to produce fertile offspring. The species is maintained by reproductive barriers between different species.*

2 a.

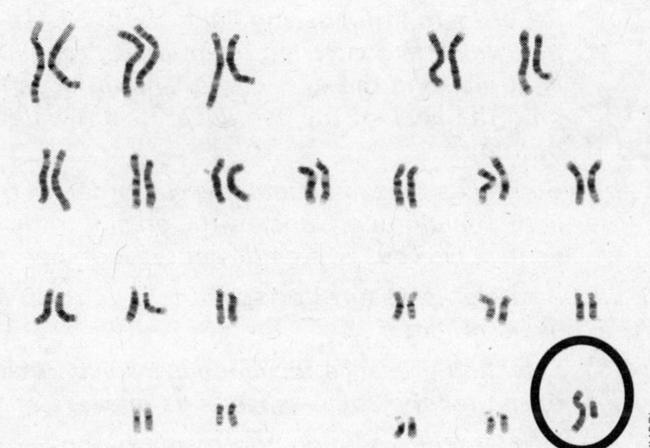

b. *Male*

c. *Normal*

d. *All chromosomes are paired. There are no aneuploidies. Chromosome number is 2N.*

3 a. *A possible dichotomous key (student answers may vary):*
1a Length of all sides equal.....A
1b Length of sides not equal 2
2a Object has a distinct tail 3
2b Object has no tail.....5
3a Object has circular hole near curved end.....B
3b Object has no hole.....4
4a Object has tail split into two lobes....:F
4b Object's tail is solid.....D
5a Object has three straight sides and one end curved....C
5b All sides are straight.....E

b.

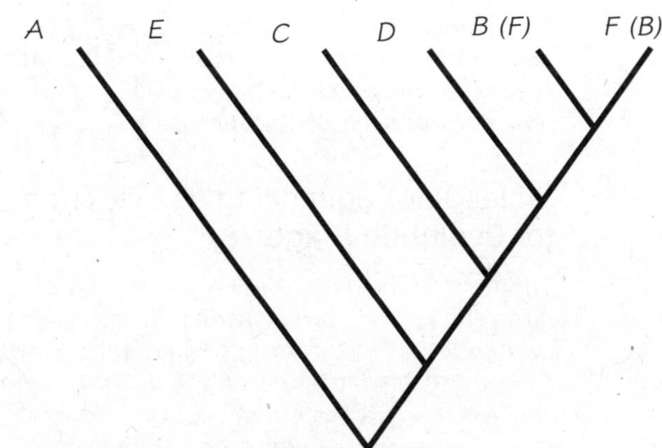

4. *Domain, kingdom, phylum, class, order, family, genus, species.*

5. *Student answers will vary, but classification may include by type (e.g. pencils, pens, erasers), by size, by colour, or by hardness.*

Theme A: Unity and Diversity

Chapter 4: Ecosystems

47. The Theory of Evolution (Page 76)

1. *Scientists had limited access to dinosaur fossils, or complete fossils, to use as evidence. Few dinosaur fossils had been discovered, or recognized for what they were. They had to base their idea about morphology of the dinosaurs on live reptiles, such as lizards and crocodiles.*

2. *Complete fossils that show how the bones were arranged. Fossils that include further evidence of skin and feathers*

3. *The variation in neck length is due to variation in genes (different alleles). If long-necked giraffes are favoured for survival and produce offspring, then their alleles for long neck are likely to be passed on with a greater frequency to their offspring.*

4. *Lamarck was one of the first scientists who attributed changes in species over time to 'natural causes'; that is, processes that were occurring in the natural world for which there was a scientific explanation. At the time, many believed that all species were formed and remained unchanged over time (immutable).*

5. *Darwin was able to observe a great many different kinds of organisms from different locations and environments around the world. He noticed patterns in their differences and similarities, linking them to common ancestry. He took extensive notes and brought back samples to investigate further.*

ISBN: 978-1-99-101423-8

6. *A scientific theory is a highly reliable explanation of natural phenomena, backed with evidence. It can still be refuted if new evidence emerges to the contrary. Most areas in science are supported by one or more theories. However, In everyday language, a theory is synonymous with having an idea, not necessarily supported by any evidence. Evidence about how inheritance functions meant that Lamarck's theory, based on traits gained in a lifetime, was discarded. This evidence was still able to fit within Darwin's observations that formed the foundation of his theory - it provided the mechanism to explain how natural selection occurred.*

48. Molecular Sequencing as Evidence for Evolution (Page 78)

1. *The level of similarity between the DNA of different species can provide information on relatedness. More closely related species will have more similarity than more distantly related species.*

2 a. *Species 1 and 3 are (9 differences).*

b. *Species 2 and 3*

3. *Humans and rhesus monkeys are both primates. They are more closely related to each other than either is to a horse, which is a mammal but not a primate. The rhesus monkey has only 2 amino differences compared to humans, whereas the horse has 25 differences. Therefore, the horse shared a common ancestor with the human much further back in time than the common ancestor between the human and the rhesus monkey. The differences in amino acids sequences can be used as evidence to indicate relatedness.*

4. *Traditional taxonomy uses anatomical homology, especially in fossils, as evidence for the extent of relatedness due to evolution. Factors, such as convergence or differences due to age or sex, can make the linking of species, or identification of species more difficult. Sequencing data is a reliable method of gauging relatedness between species, as the background mutation rate provides an accurate measure of change over time.*

49. Evolution and Selective Breeding (Page 80)

1 a. *Humans act as selection pressure.*

b. *Selective breeding is able to produce specific, useful traits in an organism that can be reliably passed on. Selective breeding is often a quicker process than natural selection. It allows us to produce offspring with desired traits over several generations quickly.*

2. *The same mechanism of inheritance that operates for natural selection to pass on traits also works for selective breeding. In selective breeding, humans drive the selection pressure to determine which individuals carrying desirable traits breed. This causes the frequency of alleles for more advantageous traits to increase in frequency in the gene pool over time.*

3. *The populations of the wild type Brassica would have shown variation in traits for leafs, stem, terminal buds etc. The growers could then select from the population which individuals to breed. For example, to develop Brussels sprouts, the grower would look for plants with larger lateral buds.*

4. *There may be some speciation, but the plants that survive and breed (through natural selection) would need to be those that are best adapted. Growers make a selection of traits, such as a large flower in cauliflower, due to their own preferences for a human crop to eat, not necessarily to survive in the wild. The larger flower (as associated alleles) would not likely appear in the wild as the trait would not be selected for.*

50. Homologous Structures (Page 82)

b. *Modified for tree climbing (flexible joints) and improved dexterity of fingers.*

c. *Modified to increase surface area and streamlined to function as a paddle.*

d. *Modified for swift running in pursuit of prey. Walks on toes, long limbs for long stride.*

e. *Short, strong limb. Shovel-like paw with sharp claws for digging and propulsion underground.*

f. *Modified into a wing for flying. Very long metacarpals and fingers stretch the skin into a wing.*

2. *The limbs all share the same basic bone anatomy, although highly modified in some cases. It is possible to match bone for bone but, at the same time, recognise how individual bones or bone groups have changed to better perform a new function for the animal.*

3. *Homology is the study of similarities present in individuals, i.e. anatomical or molecular, that are evidence for evolution. Fossils of whale species would show homology with species that are ancestral, as well as species that are more recent or living. The degree of difference would allow placement with closest relatives.*

51. Convergent Evolution (Page 83)

1 a. *Receiving and focussing light - both eyes have a gap which is controlled by an iris to regulate the light entering the eye, and a lens to focus light onto the back of the eye. Operates in water or in air.*

b. *Both eyes have a retinal layer in order to receive light stimuli (and transmit the stimuli to the central nervous system/brain through nerves).*

2 a. *Selection pressure for speed to catch fish and avoid predators.*

b. *Selection pressure for manoeuvrability (quick turns) for the same reasons as above.*

3. *Convergent evolution can result in analogous structures which appear to be similar, e.g. body shape, eye, but have quite separate evolutionary origins - some features, like eyes/sight, or flightlessness in birds, can evolve multiple times. Additional molecular homology can provide a more accurate picture of relationships.*

ISBN: 978-1-99-101423-8

52. Speciation (Page 84)

1. *The scientific name of the two types of gorilla indicates that they are not separate species, but instead sub-species that can still (likely) breed with each other. Not enough time has passed for full speciation to occur between the mountain and Eastern lowland gorilla.*
2. *Speciation is due to selection pressure on an existing gene pool (frequency/type alleles), and highly unlikely those same alleles will be present (due to extinction), therefore the same species can not be recreated.*
3. *The two species have not diverged far enough from each other currently (this could be due to time, or possible unknown gene flow). In the wild, the river is a very effective geological barrier, preventing breeding. Species that occupy the same location need to have additional forms of isolating mechanisms (reproductive) to prevent interbreeding, and allow speciation to occur, otherwise species would hybridize.*
4. *Reproductive isolation (either post- or pre-zygotic) would allow species to occupy the same location, but prevent interbreeding/gene flow between populations of different species. The genetic differences between the species, due to mutation / natural selection, creating differences in behaviour, physical traits etc., would be too great to allow successful breeding and/or viable offspring.*
5. *Speciation is the formation of new species, and this can occur when differential selection pressures are applied to separate populations - leading to a change in gene pools between species. Isolating mechanisms, either geological or reproductive, prevent gene flow between populations - which can prevent speciation.*

53. Allopatric and Sympatric Speciation (Page 86)

1.

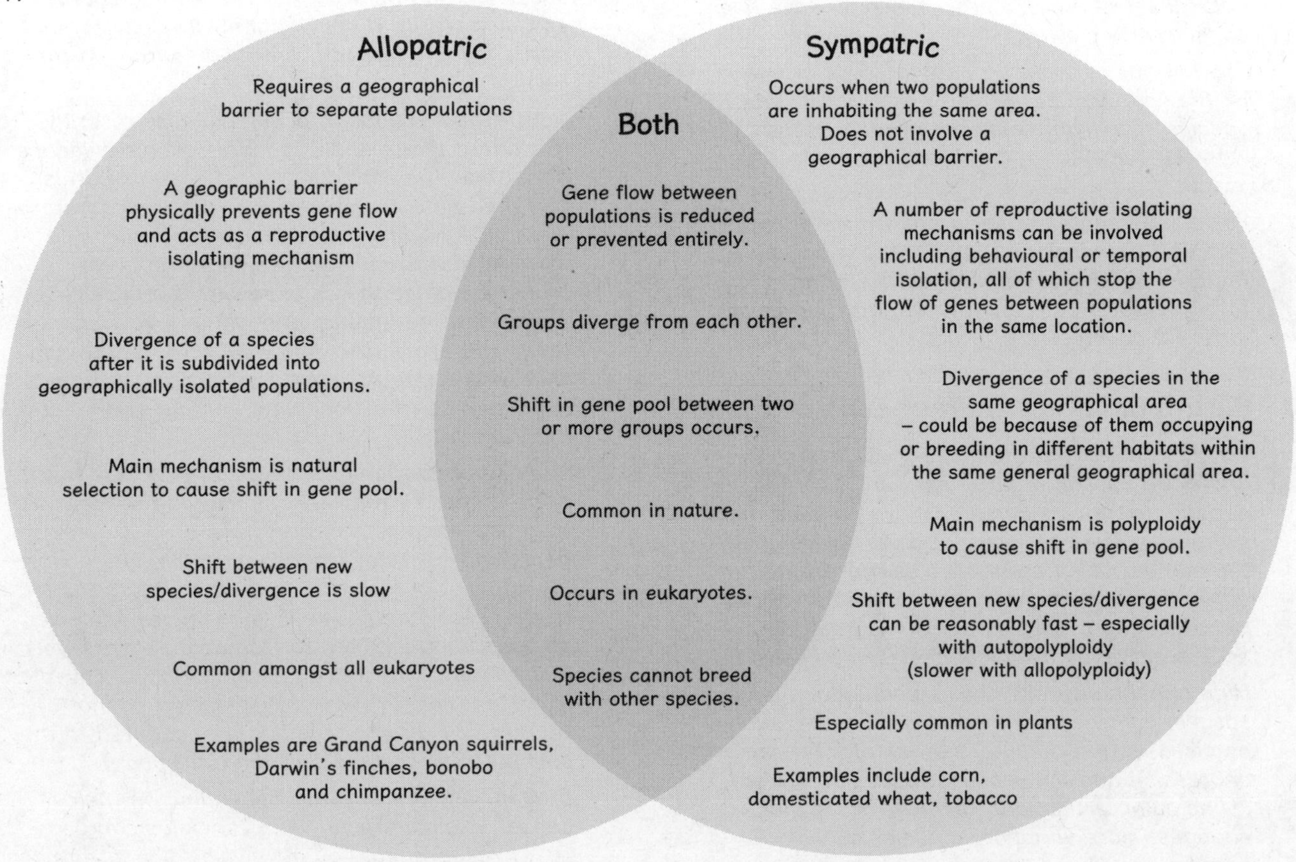

Allopatric

Requires a geographical barrier to separate populations

A geographic barrier physically prevents gene flow and acts as a reproductive isolating mechanism

Divergence of a species after it is subdivided into geographically isolated populations.

Main mechanism is natural selection to cause shift in gene pool.

Shift between new species/divergence is slow

Common amongst all eukaryotes

Examples are Grand Canyon squirrels, Darwin's finches, bonobo and chimpanzee.

Both

Gene flow between populations is reduced or prevented entirely.

Groups diverge from each other.

Shift in gene pool between two or more groups occurs.

Common in nature.

Occurs in eukaryotes.

Species cannot breed with other species.

Sympatric

Occurs when two populations are inhabiting the same area. Does not involve a geographical barrier.

A number of reproductive isolating mechanisms can be involved including behavioural or temporal isolation, all of which stop the flow of genes between populations in the same location.

Divergence of a species in the same geographical area – could be because of them occupying or breeding in different habitats within the same general geographical area.

Main mechanism is polyploidy to cause shift in gene pool.

Shift between new species/divergence can be reasonably fast – especially with autopolyploidy (slower with allopolyploidy)

Especially common in plants

Examples include corn, domesticated wheat, tobacco

54. Adaptive Radiation in Mammals (Page 87)

1. *With the extinction of the non-avian dinosaurs and aquatic reptiles, many niches were made vacant, providing the opportunity for mammals to diversify rapidly to occupy them. This radiation resulted in an increase in diversity and morphological change so that mammals came to occupy all the niches previously occupied by non-avian dinosaurs.*
2. *Common ancestors.*
3. *A divergence occurred during the Jurassic period (one group of early mammals split into two lines: marsupials and placentals).*
4. *They were ancestral to many other mammal orders. They shared a common ancestor in the Cretaceous period.*
5. *Rodents and odd-toed ungulates (widest green shape which indicates the largest number of species).*
6. *Their clade became extinct and they have no more living representatives on earth.*
7. *Paleocene (66-53 mya).*

ISBN: 978-1-99-101423-8

55. Barriers to Hybridization (Page 88)

1. *Prezygotic isolating mechanisms prevent or reduce the chance of mating occurring. Should mating occur, then the post zygotic isolating mechanisms prevent a viable offspring being produced or, if offspring are possible, then they will be infertile and unable to breed, This reduces or prevents gene flow between species.*

2 a. *Viable offspring cannot be formed, either by death or embryo, or soon after birth.*

b. *Offspring can occur - but they will be unable to breed - often through chromosome mismatch.*

c. *The 1st generation offspring can breed, but the resulting 2nd generation is infertile or dies.*

56. Abrupt Speciation in Plants (Page 89)

1. *The giant and Japanese knotweed are two separate species. Polyploidy occurred in giant knotweed instantly, the result being a new species, Japanese knotweed - with double the chromosomes, and different traits. If the different traits / speciation occurred due to natural selection, this process would have been considerably slower.*

2. *The hybrids may not function as species - that is, they can still easily breed with other 'parent' species. The hybrid may not produce offspring with the same set of traits, or be able to breed at all.*

3. *Polyploidy creates offspring that have multiple sets of chromosomes compared to a parent (mutation), whereas allopolyploidy is the formation of a polyploid by hybridization from different species.*

57. Earth's Biodiversity (Page 90)

1. *Genetic diversity is the number of genetic characteristics in a particular species. High genetic diversity is beneficial to a population as it makes it more resilient. Species diversity is the number of different species represented in a particular community (also known as species richness. High species diversity is healthy for an ecosystem as all available niches are filled.*

2. *High genetic diversity makes populations less susceptible to environmental challenges, such as disease, and reduces the risk of inbreeding depression and the problems associated with it (including extinction). If the environment changes, more variation in alleles in the population will make it more likely that one or more individuals are able to survive the changed abiotic/biotic conditions, and produce offspring (i.e. not become extinct).*

3. *The environment in the water may have been more stable, or been able to supply organisms with nutrients that were not available on land, due to lack of plants (to feed on and recycle nutrients (nitrogen)). Organisms may not have evolved adaptations so they could support themselves without water buoyancy or maintain temperature.*

4. *Science is not static. Different hypotheses facilitate research and potentially new evidence being uncovered so knowledge moves in the direction of being more reliable. As both hypotheses have yet to be discarded, ongoing research to find new evidence for both can still continue.*

58. The Sixth Mass Extinction (Page 92)

1. *The Sixth Extinction is the human-induced loss of much of the Earth's biodiversity, measured by an extinction rate well above the 'normal' background extinction rate.*

2 a. *17.4 extinctions per century. (5487/1,000,000) x 100 = 0.5487. 17.4/0.5487 = 31.7 times greater.*

b. *4.4 extinctions per century. (10,000/1,000,000) x 100 = 1. 22/1 = 22 times greater.*

c. *7.8 extinctions per century. (6700/1,000,000) x 100 = 0.67. 7.8/0.67 = 11.6 times greater.*

d. *22.8 extinctions per century. (300,000/1,000,000) x 100 = 30. 22.8/30 = 0.76 of rate.*

3. *When they move into an area, humans can introduce pests that can directly compete with species. The human requirement for resources can also result in competition with species, in particular the habitat, and food sources in the habitat.*

4. *As the moa extinction occurred more recently, in modern times, there is more reliable evidence about how the extinction event occurred. This can then be extrapolated onto other megafauna extinction events, where these species came in contact with humans. Key features were that the extinction occurred with a low density/ population of humans, and within a very short time (150 years) the moa went from healthy and widespread populations to extinction. Multiple impacts occurred, including hunting and habitat destruction.*

5. *The Caribbean monk seal was one of only three related species that were unique in the fact that they had adapted to warm waters. They were genetically distinct from other seals, and shared a common ancestor not long after the formation of the 'seal' clade. The loss of this seal species also means the genes for the unique traits were lost as well. The seal species may have played an important role in the ecosystem that cannot be replaced by another species (not adapted to the particular environment of the Caribbean).*

6. *Student answer will vary depending on their chosen example. If no local examples can be easily researched, you may wish to choose one of the following examples:*
    - *Tasmanian tiger / thylacine (Thylacinus cynocephalus)*
    - *Western Black Rhino (Diceros bicornis ssp. longipes)*
    - *Pyrenean Ibex (Capra pyrenaica pyrenaica)*
    - *Spix's Macaw (Cyanopsitta spixii)*
    - *Splendid Poison Frog (Oophaga speciosa)*

    Format should follow the two case studies provided in the worktext. The Tasmanian tiger is provided as an additional example.

    Thylacine (Thylacinus cynocephalus) also called

ISBN: 978-1-99-101423-8

the Tasmania tiger or Tasmanian wolf, was a carnivorous 'dog-like' marsupial.

Species origin: The Thylacinus genus first appeared in the fossil record around 16 mya, and fossils of the Tasmanian tiger fossil record dates back to 1.7-0.78 mya in Australia.

Species' previous distribution and population number: Thylacines were once widespread across the Australian mainland, Tasmania, and New Guinea but were thought to have become extinct on the Australian mainland and New Guinea ~3600-2000 years ago. There are no remaining populations in Tasmania and the last animal in captivity died in 1936. The species is now extinct.

Causes of extinction linked to humans: Human hunting (especially between 1830 and 1920) as it was deemed a pest, competition from domestic dogs (and diseases they introduced), and habitat destruction. Competition with dingoes may also have contributed to its extinction.

59. Human Activity and Ecosystem Loss (Page 94)

1. *The rate of loss is much higher than would be expected by natural events. Most of this ecosystem loss is identified as being caused by humans. Human demand for resources continues to increase, often in competition with other species in ecosystems. This puts extra stress on the ecosystems and the species in them, which could lead to their collapse.*
2. *The forests provide industry and employment at an economic level and at a human level, including food such as fruits and meat. This makes is difficult to convince people to conserve parts of the forest as they lose access to these parts and so lose potential resources.*
3. *Student answer will vary depending on their chosen example. Some possible examples could include:*

   Caribbean coral reefs, Alaskan kelp forest, Aral sea, The Great Barrier Reef.

   The Alaskan kelp forest has been provided as an example.

   Location: Rocky coastlines along the Pacific (Western) coast of Alaska, particularly the western Gulf of Alaska, the eastern Bering Sea, and the Aleutian Islands.

   History: Kelp is a large, brown algae. There are around 30 species but bull kelp and giant kelp are important species in the Alaskan kelp forests. Kelp grow in cool, high nutrient, shallow, coastal waters and grow rapidly in good conditions, e.g. bull kelp can grow 25 cm /day.

   Key aspects: Provide a unique marine ecosystem and form habitat for a large number of crustaceans, sea snails, sea urchins, starfish, fish species and sea otters. The kelp forests provide food, shelter and breeding grounds for many species, including commercially important species. They are also important photosynthesizers. Kelp forests can become depleted by overgrazing from sea urchins and fish species, but also human activity. Healthy populations of sea otters and sea stars prey on the fish and sea urchins to keep their numbers in check and this minimizes damage to the kelp forests.

   Human activity: Many human factors contribute to reduction in the kelp forests. These include kelp harvesting, pollution, e.g. oil spills and climate change which can alter the ocean temperature affecting reproduction and survival of the kelp species.

   Consequences: Reduction of the kelp forests puts the many species that depend on them at risk, potentially resulting in lost biodiversity and reduced ecosystem stability. E.g. loss of keystone species, the sea otter, which could result in large numbers of sea urchins which over-graze the kelp. The ecosystem could collapse. Collapse of ecosystem. Loss of commercial fish stocks would impact biodiversity and human food supplies and income.
4. *Statements about the biodiversity crisis to encourage general public assistance or policy change, or generate funding, may not be believed without sufficient evidence for quality research. One-off research projects or sets of data have less reliability, as results may be due to a chance occurrence or a poorly representative sample. Increasing the number of repeats also increases data reliability, verified with statistical tests.*
5. *Citizen science has the potential to 'recruit' an extremely large volunteer group to collect data, often with local knowledge, and not requiring extensive funding. The expertise of the volunteers may be variable, especially those without correct training. Also, the reliability of the data may be reduced, and more bias could be included in the collection method.*
6. *Student answer.*
7. *There are multiple ways that humans can contribute to the biodiversity crisis and in some ecosystems one impact will lead onto another. As the human population increases, along with resource demand and waste production, the impacts will be more likely to increase. Wild land is deforested and used for either food production (agriculture/monoculture) or urbanization. Species which once occupied these ecosystems may be found nowhere else (endemic) and could become extinct. Disease and pests can be spread across different countries, outcompeting and killing native species. For example.... (local example).*
8. *Student answer will vary depending on their chosen example.*

   Students may choose a high profile global issue, e.g. destruction of the Amazon rainforest to make way for agriculture, or a local example such as the development of land for new housing or the damming of a river for hydroelectric energy production.

   The report should identify:

   - *The area and type of human activity affecting it, e.g. deforestation, pollution, flooding etc.*
   - *Why the area is/was important for*

ISBN: 978-1-99-101423-8

biodiversity, identifying any unique or keystone species or factors if applicable.

- Identify the consequences of biodiversity loss. For example, destruction of a unique habitat, loss of keystone species, ecosystem collapse, loss of unique species, potential loss of organisms which can contribute to new medicines or technology, loss of carbon sinks etc.
- Students may include information on mitigation steps to minimize biodiversity loss associated with their case study.

60. Conservation Strategies (Page 97)

1. *In-situ conservation uses ecosystem management and legislation to protect species in their natural habitat. By necessity, this involves both restoring the ecosystem and implementing laws to protect the species of interest. Methods include protecting and/or restoring the habitat, and protecting the endangered species from predators, hunting, and illegal trade. Neither tool is effective in isolation; if the species are not protected, there is little point in restoring their habitat and, if they are without habitat, the species would have to remain in a zoo.*

2. *Ex situ conservation methods are often employed when species numbers become critically low or in situ methods are not working. Features of ex situ conservation focus on (1) removal of the species from its natural habitat to a new location, usually a protected /controlled area, compared to leaving species in their original habitats with in situ methods and (2) captive breeding (or cultivation) involving a managed breeding (seed) register to maximise remaining genetic diversity. In situ conservation is focussed on increasing breeding, but mainly by natural means. Both methods can be expensive, depending on size of operation.*

3. *Humans tend to have a natural affinity for 'cute and cuddly' species (e.g. meerkats, pandas, seals, and big cats) and give them a larger share of funding. Animals such as spiders and snakes tend to produce an aversion reaction in humans. We therefore tend to be less invested in 'unattractive' creatures than the attractive ones. This affects the species' survival, if they are endangered, due to less willingness to fund conservation efforts.*

4. *Student's evaluation: they will most likely have a subjective opinion and may state that all species that are endangered deserve equal attention. Some students may acknowledge that all species cannot be saved, so a triage approach must be used, with more visible species being used to get attention for others.*

5. *The student debate should address the specific ethical, environmental, political, social, cultural, and financial consequences associated with protecting/not protecting their chosen endangered species.*

61. Did You Get It? (Page 100)

1 a. *It can identify the closeness of relationships between species, even when identification based on anatomy may be difficult. It is a reliable method that is easily verifiable.*

b. *Very few fossils have non-damaged DNA samples available and therefore molecular sequencing cannot occur.*

c. *Any different, new fossil is assigned as a new species in the splitting approach. The lumping approach acknowledges that age/sex and natural variation can cause differences, and groups fossils as one species.*

2 a. *Traits (physical or behavioural) are selected for in adults, selected by humans, and bred to increase the number of individuals with the trait.*

b. *Traits are coded into genetic material, and when selected for, and bred to increase the number of offspring, the trait (and alleles) increase in a population.*

c. *The distance and intervening water acts as a geological barrier between mainland and island populations and prevents gene flow between species. Eventually, the differences may be so great between the birds that this acts as a reproductive isolating mechanism - i.e. changes in behaviour or pre- / post-zygotic barriers.*

3 a. *Ex-situ conservation methods are often used when species numbers become critically low or in-situ methods are not working or are not feasible. Ex-situ conservation involves removing the endangered species from its natural habitat to a new location where it can be monitored and more easily protected.*

b. *Genetic diversity is important for the health of a species. It makes it more resilient to change in the environment, where some individuals may have the alleles/traits/adaptations to survive and breed. It also increases the health of the individuals - inbred organisms are more susceptible to disease.*

62. Summary Assessment (Page 101)

1. *(d) Convergent evolution*

2. *(b) C-G, T-A*

3. *(d) Conditions for the origin of life*

4. *(a) Mitochondria*

5. *(d) 46 & 48*

6. *(a) Anthropogenic impacts*

7 a. *Eukaryotes have membrane encased organelles - including mitochondria, and a nucleus. Prokaryote cells do not have mitochondria or a nucleus (the genetic material is free floating in the cytoplasm.*

b. *LUCA: Last Universal Common Ancestor. A hypothesis that a unicellular organism called LUCA (around 4 billion years ago) was the ancestor of all living organisms on earth. All organisms alive today descend from this organism (form one clade).*

c. *The sugar/phosphate forms an alternating backbone for the double helix (DNA). The bases connect as rungs across the centre, with*

ISBN: 978-1-99-101423-8

hydrogen bonds (C-G and T-A in DNA and C-G and U-A in RNA). The order of the bases codes for expressed protein (in codons).

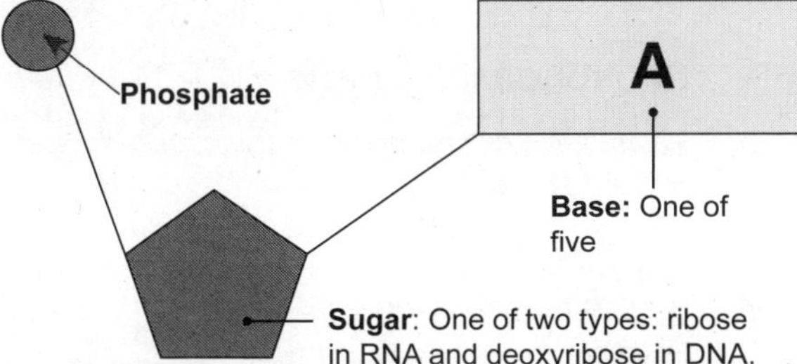

8 a. *There may be many common names, but only one scientific name: genus and species. The name also gives an indication of closely related species.*

b. *The butterflyfish all have one common ancestor from which they have descended - so form one clade. Further back in time, an ancestor including the butterflyfish and anthias was present, so they and all other species form a clade.*

c. *The chromosome number has been formed by the joining and separation of chromosomes during meiosis (and chromosome mutations, such as polyploidy) in the past generations/ancestors - and the number does not indicate complexity or the exact nature of relatedness. i.e. the two fish are closer related than to the coral.*

d. *The commonality on amino acids in a sequence (in the same gene of the three species) indicate closer relatedness - therefore they would have a common ancestor more recently than that of all three.*

9. *The most recent mass extinction is called the sixth mass extinction. Whereas the other extinction events were due to natural causes, the 6th extinction is almost exclusively due to anthropogenic (human) activity.*

10. *In-situ conservation protects the species in an ecosystem within their natural habitat. This may mean restoring an ecosystem, or removing pest species. The coral reef system is complex, and would involve the translocation of many species together, in the correct environment, in order for it to be stable enough to be successful.*

11 a. *Divergent - many different species that are related (a common ancestor in the Rabdophorus group) speciating out in different habitats. They look similar but not the same. (i.e. same shape - different colours).*

b. *It was likely that the three species had a common ancestor that lived in the Caribbean. At some stage two populations diverged (probably with a geographic isolating mechanism to stop gene flow) - one population migrated to the colder Atlantic, adapting to the cooler water (C. ocellatus). The other population diverged again at a later date - to form one species (C. striatus) that also migrated to the Atlantic, and the another (C. capistratus) that remained in the warmer Caribbean.*

Theme B: Form and Function

Chapter 5: Molecules

63. Carbon Chemistry (Page 105)

1. *The ability of carbon atoms to form four covalent bonds.*
2. *Bonds 3, 4, 5, 6*
3.

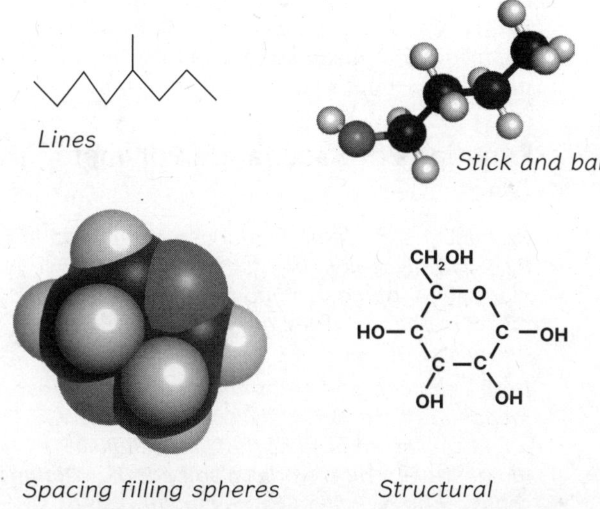

64. Carbohydrate Chemistry (Page 106)

1 a. *Primary energy source for cellular metabolism*

b. *Structural units for disaccharides and polysaccharides (energy sources and structural carbohydrates).*

2. *Glucose is a hexose sugar (6 carbon atoms). Ribose is a pentose sugar (5 carbon atoms).*

3. *Isomers have the same molecular formula but their atoms are linked in different sequences. α-glucose and β-glucose are isomers because although they have the same molecular formula, they are structurally different and have different properties.*

65. Condensation and Hydrolysis of Sugars (Page 107)

1. *Disaccharide sugars are formed by condensation) and broken down by hydrolysis. Condensation reactions join two monosaccharide molecules by a glycosidic bond and release a water molecule. Hydrolysis reactions use water to split a disaccharide molecule into two monosaccharides. The water molecule provides a hydrogen atom and a hydroxyl group.*

2. *A : Reaction = Condensation. Product: =maltose.*

   B : Reaction = Hydrolysis. Product= two glucose molecules.

3. *The monomer involved and its isomer (alpha or beta).*

66. Polysaccharides (Page 108)

1 a. *They are easily hydrolysed into monosaccharides (e.g. glucose) when energy is needed. Monosaccharides are the primary source of cellular fuel.*

ISBN: 978-1-99-101423-8

b. Polysaccharides are hydrolysed to produce simpler carbohydrates, e.g. glucose, which can then be transported to other parts of the organism.

2. Glycogen is a highly branched glucose polymer. It is compact and easily hydrolysed to provide glucose for energy. Glycogen can be metabolized quickly, which suits the active lives of animals. Starch is a mix of branched and unbranched chains of glucose. It is powdery and less compact than glycogen, insoluble in cold water, but relatively easy to hydrolyse to soluble sugars, making it a good storage molecule for plants. Cellulose is a linear glucose polymer and is strong and insoluble, making it well suited to provide strength and support to plant cells.

67. Functions of Saccharide Polymers (Page 109)

1. Cellulose is made of β glucose monomers joined by β 1,4 glycosidic bonds, whereas amylose is made of α glucose monomers joined by α 1,4 glycosidicbonds. They are both long, linear polymers.

2. Both cellulose and chitin are made of glucose monomers with β 1,4 glycosidic bonds. We could therefore predict that, as cellulose is indigestible, chitin would also be indigestible.(In reality, chitin is slightly digestible).

3. A ends with a galactose molecule, B with an N acetyl-galactosamine molecule and O is missing the terminal molecule.

68. Lipids (Page 110)

1 a. Glycerol

b. Ester bond

c. Fatty acid

2. Fats need more oxidation per gramme to form CO_2 and water than other molecules (the carbon atoms in fatty acids have more electrons around them) and so more usable energy can be extracted from the molecule.

3 a. Saturated fatty acids have the maximum number of hydrogen atoms, whereas unsaturated fatty acids contain some double-bonded carbon atoms.

b. Neutral fats with a high proportion of saturated fatty acids tend to be solid at room temperature. Neutral fats with a high proportion of unsaturated fatty acids tend to be liquid at room temperature.

4 a. During esterification, a glycerol molecule is joined with a fatty acid. This occurs three times to form a triglyceride.

b. Hydrolysis of a triglyceride produces glycerol and three fatty acids.

5. Lipids are a more concentrated source of energy than carbohydrates or proteins, providing fuel for aerobic respiration through fatty acid oxidation. They are important as energy storage molecules. Fat absorbs shocks and cushions internal organs. Stored lipids provide insulation and reduce heat loss. Lipids are a source of metabolic water (fat stores can be metabolized to provide water as well as energy). As steroids, they are important as hormones and transport fat soluble vitamins. Waxes and oils provide waterproofing to the surfaces of organisms and phospholipids form cellular membranes.

69. Phospholipids (Page 112)

1 a. Triglycerides comprise a glycerol attached to three fatty acids via ester bonds. Phospholipids have the same glycerol base except the final fatty acid is replaced by a phosphate group and a nitrogen-containing compound.

b. Phospholipids are amphipathic: they have a polar hydrophilic end and a hydrophobic fatty acid end. This structure causes them to orient in aqueous solutions so that the hydrophobic 'tails' point in together. This is important in their functional role forming the bilayer structure of membranes.

2. Unsaturated phospholipids cannot pack together as tightly as saturated phospholipids. There are more spaces within the membrane bilayer, making the membrane more fluid.

70. Crossing the Lipid Bilayer (Page 113)

1 a. Gases (CO_2, O_2), hydrophobic molecules such as benzene, small polar molecules such as ethanol and H_2O

b. Charged molecules such as ions e.g. Na^+, Ca^{2+}, Cl^-

c. Large polar molecules, e.g. glucose

2. Steroids are non-polar and are therefore able to dissolve into the hydrophobic tails of the phospholipids that make up the lipid bilayer. They can then pass through the bilayer.

71. Amino Acids (Page 114)

1 a. Amino acids have the same general structure: an amine group, carboxyl group, hydrogen atom, carbon atom, and an R group (a variable side group).

b. The side chains (R groups) differ in their chemical structure (and therefore their chemical effect).

2. Twenty different amino acids comprise the building blocks for constructing proteins (which have diverse structural and metabolic functions). The non-protein amino acids have specialized roles as intermediates in metabolic reactions or as the precursors of many important molecules. Amino acids are also available as dietary supplements for specific purposes.

72. Amino Acids and Proteins (Page 115)

1. R groups drive the folding of each polypeptide chain and hold the separate chains together, maintaining the functional structure. In insulin, the two chains are held together by disulfide bridges between cysteines.

2. The interior of a plasma membrane is a hydrophobic environment. Channel proteins span the membrane and fold in such a way that the non-polar (hydrophobic) R-groups align to the outside, and polar (hydrophilic) R-groups form a channel on the inside. This channel allows water soluble molecules to cross the membrane.

ISBN: 978-1-99-101423-8

73. R-Groups (Page 116)

1 a. *Ionic bond, hydrophobic interactions, disulfide bonds, hydrogen bond.*

b. *Disulfide bonds and ionic bonds.*

2 a. *Gly, Ala, Val, Met, Leu, Ile, Phe, Tyr, Trp*

b. *Ser, Thr, Cys, Pro, Asn, Gln*

c. *Lys, Arg, His*

d. *Asp, Glu*

3. *Polar and charged amino acids would be on the surface because they can interact with water. Nonpolar amino acids would be on the inside because they are hydrophobic and will be kept away from the water.*

74. Protein Structure (Page 117)

1 a. *Peptide bonds between the amino acids.*

b. *2° structures (e.g. α-helices and β-pleated sheets) form as a result of hydrogen bonding between neighbouring CO and NH groups.*

c. *Chemical bonds (e.g.disulfide bridges), maintain 3° structure. These interactions arise as a result of the different properties of the amino acid R groups*

d. *Interactions between two or more polypeptide chains produce the functional protein. Bond interactions are as for maintenance of 3° structure.*

2. *The R groups on the amino acids allow weak intermolecular forces to bind different parts of the polypeptide together. This causes the polypeptide to fold into its functional shape.*

3. *A conjugated protein has a non-amino acid group attach to it (the prosthetic group). Non-conjugated proteins are comprised of only amino acid residues.*

4. *The development of computer technology has helped in determining protein structure by being able to analyse much more information than was previously possible. Both x-ray crystallography and cryogenic electron microscopy use computers to analyse images. AI can predict protein shape for any sequence of amino acids.*

75. Comparing Globular and Fibrous Proteins (Page 119)

1. *Many globular proteins are enzymes involved in catalysing metabolic reactions in cells, e.g. RuBisCo catalyses the first step of carbon fixation in photosynthesis. Some globular proteins have a regulatory role as hormones (e.g. insulin) and others have a transport role, carrying specific molecules from one site to another (e.g. haemoglobin transports oxygen). Antibodies (immunoglobulins) have a role in internal defence. A few globular proteins (not many) have a structural role as monomers contributing the cytoskeleton of cells.*

2 a. *The tertiary structure of globular proteins provides a specific 3-dimensional shape that allows them to interact with other molecules and perform their function. For example, haemoglobin has a complex quaternary structure incorporating 4 haem groups that bind oxygen molecules for transport.*

b. *The tertiary structure is critical to function so a change in tertiary structure would lead to loss of functionality. The protein would lose its ability to perform its biological role.*

3. *Fibrous proteins, such as collagen and elastin, are the major component of many connective tissues, including tendons and ligaments (and also in skin), providing support and rigidity to the more fluid components of tissues. Keratins are fibrous proteins that make up hair, nails, wool, feathers, horns, and hooves and are important in forming durable structural and functional components of organisms.*

4. *The tertiary structure of fibrous proteins produces long fibres or sheets, with many cross-linkages. This makes them very tough physically and ideal as structural molecules. For example, collagen consists of polypeptides wound together to form rope like structures, which then self assemble into fibrils held together by covalent cross linkages.*

5. *All three fibrous proteins form stable, covalent cross linkages between amino acid residues in adjacent polypeptide chains, making stable and strong fibrous structures.*

76. Did You Get It? (Page 121)

1 a. *The hydrophilic end of the phospholipid molecule.*

b. *The hydrophobic end of the phospholipid molecule.*

c. *The hydrophilic end orients towards water, the hydrophobic end orients away from water. This results in the formation of the bilayer with hydrophobic ends in the interior and hydrophilic ends on the (aqueous) exterior.*

2 a. *Proteins. There are alpha helices visible which are characteristic of proteins.*

b. *Increased temperature, large changes in pH.*

c. *The functionality would be lost.*

d. *Quaternary (there are four complete polypeptide chains in the protein).*

3 a. *Condensation (dehydration synthesis).*

b. *Hydrolysis*

c. *Water is released during a condensation reaction. In a hydrolysis reaction water is incorporated into the molecule.*

4 a. *R-group*

b. *Amine residue*

c. *Peptide bond*

5. *The protein is a conjugated protein. The prosthetic group can be seen to the left of the main chain.*

ISBN: 978-1-99-101423-8

Theme B: Form and Function

Chapter 6: Cells

77. The Plasma Membrane (Page 124)

1. *Phospholipids, proteins, cholesterol.*

2 a. *Water, lipid soluble molecules (e.g. carbon dioxide), non-polar steroids.*

b. *Channel protein.*

c. *Sugar, amino acids, and nucleosides (nucleic acid components).*

3 a.

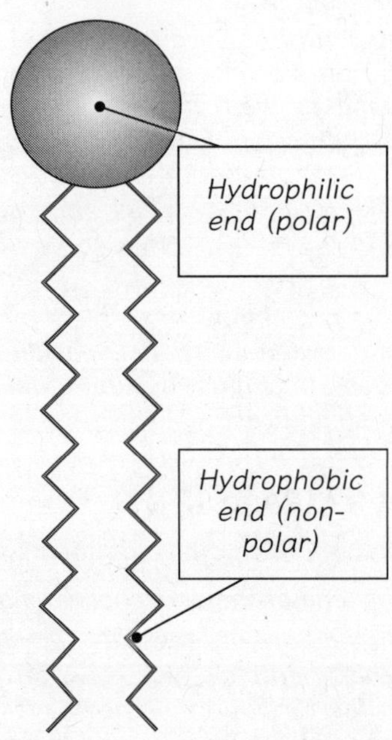

b. *The bilayer structure, with its integrated proteins allows only certain molecules to pass through.*

4 a. *The hydrophobic tails of the phospholipids repel and prevent polar and charged molecules passing through. Water requires a concentration gradient to move using osmosis. Large particles don't fit through the layer.*

b. *Diffusion of simple gases and non-polar steroids is a passive process, and the particles move freely in all directions, including through the membrane. Lipophilic (attracted to lipids) steroids are 'drawn' inside the membrane. Water moves by osmosis, so moves in one direction across a semi-permeable membrane.*

5. *Models are a simplified version of structures and processes and are tools used in science to provide clearer explanations about complex and abstract concepts. For example, the molecules involved in the membrane are not readily identifiable and many are too small to visualize from images.*

6. *This model accounts for the properties we observe in cellular membranes: its fluidity and its mosaic nature. The fluid mosaic model also accounts for how membranes can allow for the selective passage of materials.*

7.

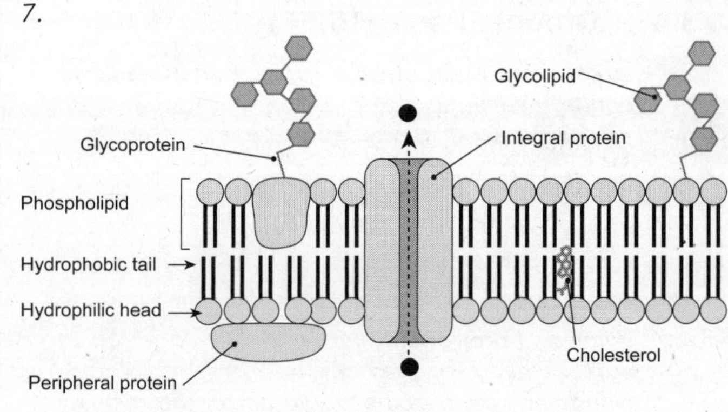

78. Proteins of the Plasma Membrane (Page 126)

1 a. *The integral proteins are found permanently embedded throughout the membrane, either entirely through (transmembrane) or partially embedded. Examples: channel proteins and carrier proteins. Peripheral proteins are only found on the outside of the membrane, and often only temporarily attached. Examples: spectrins.*

b. *Integral proteins are involved with channelling (channel proteins) or transporting molecules (carrier proteins), while some can act as cell receptors. Peripheral proteins are also involved in transport, and as 'accessory" enzymes and communication. They can form part of the cell's support - the cytoskeleton.*

2. *Both have carbohydrates that are located on the exterior surface of the membrane, but glycoproteins also have integral protein embedded in the membrane.*

3. *Glycoproteins have a carbohydrate component that is covalently bonded directly to an integral protein. Glycolipids are solely composed of a carbohydrate component that is attached directly to the lipid layer with a glycosidic covalent bond. The carbohydrate component of both molecules extends out into the exterior surface of the membrane.*

4. *Both glycoproteins and glycolipids are involved in structural support of the cell and enable connections and communication between other cells. Whereas a key role of glycoproteins is to act as receptors for chemical signals, glycolipids function to recognize the difference between their own body cells and other foreign cells. Both of these roles are crucial for the immune system in triggering immune responses. Both glycolipids and glycoproteins are involved with antibody recognition to act as antigens in ABO blood typing. This is a result of the same type of carbohydrate attached, an oligosaccharide - a molecule consisting of 3 to 10 sugar (monosaccharides) units attached together.*

79. Movement Across the Plasma Membrane (Page 128)

1. *They all involve movement of substances across a membrane with no input of energy (they are passive) with substances moving down a concentration gradient, from high to low concentration.*

ISBN: 978-1-99-101423-8

2. *Facilitated diffusion uses channels or carriers to allow ions and lipid-insoluble molecules to pass through the membrane. Simple diffusion does not, and the small molecules move directly through the lipid layers, from one side to the other. Water movement uses both methods - but aquaporins can facilitate faster movement.*
3. *Ions need to be concentrated on one or the other side of a membrane to build up resting potential. The gates need to be closed on ion channels so constant flow does not occur (and neutralizes charge). An electrical signal can be quickly transmitted when the gates then open, and ions / charge moves from one side to another.*
4. *The membrane surface of the thin, flat cell would be closer to the cell centre than for the thick round cell. Particles would be able to diffuse across the shorter distance more rapidly.*
5. *Osmosis is the diffusion of water molecules across a partially permeable membrane from a region of lower solute concentration (higher free water molecule concentration) to a region of higher solute concentration (lower free water molecule concentration).*

6 a. *Right*

b. *The water moved from left to right across the membrane because it contained more dissolved particles and therefore had a higher solute concentration and a lower concentration of free water molecules. The water moved down its concentration gradient, from high concentration (left) to lower concentration (right).*

7. *An increase in solutes - as ions - inside the cytoplasm of the plant root hair cell increases the osmotic pressure. Endosmosis will increase and be at a greater rate than exosmosis as water in the surrounding soil moves across the membrane of the cell and into the cell. This means that there is a NET flow of water by osmosis into the plant cell. This will increase the amount of water available to the plant, reducing the water demand.*
8. *Where water needs to move from one tissue/ cell to another inside the body to maintain homeostasis, the aquaporin channels increase the speed/efficiency of osmotic movement of water. For example, when the blood pressure needs to be increased in the blood vessels, water can be moved from the interstitial space, across the membrane through aquaporin channels, and into the blood vessels cell - and then into the blood.*
9. *A is showing simple diffusion of water across a membrane and into a cell using the process of osmosis. B is showing water also moving by osmosis but through an aquaporin channel to make the process more efficient / faster. C is showing facilitated diffusion of ions through a carrier protein, where the ions would otherwise be inhibited by direct diffusion across the membrane due to the polarity of the ion being repealed by the polarity of the plasma membrane.*

80. Active Transport and Pump Proteins (Page 131)

1 a. *Active transport is the energy using process of moving molecules or ions against their concentration gradient.*

b. *It moves molecules against their individual concentration gradients to transport essential molecules or ions across membranes.*

2. *ATP*
3. *Pumps use energy (from an external source) to push water against a gradient, just as pump proteins use energy (from ATP) to push molecules against a gradient. In the case of the water pump, the gradient is gravity. In the case of the pump proteins, the gradient is concentration of the molecule.*

81. Membrane Fluidity (Page 132)

1 a. *Saturated fatty acids form a straight line lipid - and these pack together tighter - making the membrane less fluid due to smaller gaps between them. Unsaturated fatty acids form a bent shape, pushing lipids away from each other to increase fluidity due to larger gaps between the lipids.*

b. *Colder environments tend to compress the phospholipids together, so some animals have evolved with adaptations to have a higher proportion of unsaturated fatty acids in their membranes to compensate.*

2. *Cholesterol increases in rigidity in warmer temperatures - providing more stability (and less fluidity) to the membrane, while in warmer temperatures it 'packs' between lipids to retain space between them for fluidity.*

82. Cytosis and Membrane Fluidity (Page 133)

1. *Phagocytosis is the engulfment of solid material by endocytosis whereas pinocytosis is the uptake of liquids or fine suspensions by endocytosis.*
2. *Ingestion of bacteria and cell debris by phagocytic white blood cells.*
3. *Secretion of substances from specialized secretory cells in multicellular organisms, e.g. export of proteins by lymphocytes.*
4. *As the phospholipid membrane is not a fixed structure, but instead fluid, parts of the membrane are able to change shape and form around substances to make a vesicle in a process called endocytosis. The vesicle can detach from the membrane and move independently around the cytoplasm. The membrane lined vesicle can rejoin with a membrane in the process of exocytosis, and expel the contents out of the cell or into an organelle.*

83. Gated Ion Channels (Page 134)

1. *To control the stopping and starting of facilitated diffusion of ions moving down the concentration gradient. This allows resting potential to build across membranes and then release the ions so an action potential can be sent, when signalled.*

ISBN: 978-1-99-101423-8

2. Nicotinic acetylcholine receptors are opened when two molecules of acetylcholine bind to the receptors. This opens the ion channel in the centre of the molecules and allows the movement of sodium and potassium ions through. Introducing nicotine to the body has the same result. Nicotine can thus stimulate nerve impulses in the same way as acetylcholine.

84. Exchange Transporters and Cotransporters (Page 135)

1. To allow for the correct concentration of sodium and potassium ions inside and outside the cell to maintain membrane potential.

2 a. Cotransport describes coupling the movement of a molecule against its concentration gradient to the diffusion of an ion (e.g. H^+ or Na^+) down its concentration gradient.

b. A Na^+ gradient established by an Na^+/K^+ pump is used to drive the transport of glucose across the epithelium. A specific membrane protein then couples the return of Na^+ down its concentration gradient to the transport of glucose.

c. Glucose diffuses from the gut epithelial cells into the blood, maintaining a low level in the epithelial cells.

3 a. Delete secondary

b. Delete primary

85. Cell-Adhesion Molecules and Junctions (Page 136)

1. A junction is the meeting point of two cells, and CAM is the type of cell-adhesion protein that holds the cells together, and sometimes forming conduits between them.

2. Connected cells can share cytoplasm (plasmodesmata / gap junctions), and ions/ small molecules (gap junctions), allowing communication between cells. Cells can be joined together to form tissue (desmosomes). Waterproof layers around specific organs can be formed by bonded cell layers (tight junctions).

3 a. In the cell wall of plant cells, and in algal cell walls.

b. In epithelial and endothelial (lining) cells in animal tissues / organs; skin, cavities, and vessels.

c. Found in nearly all animal cells that are connected together to form tissue, including the heart.

d. In animal cells to add strong attachment between cells, including cardiac muscle, bladder, epithelial.

86. Compartmentalization in Cells (Page 137)

1. – The Golgi is composed of membranous sacs specialized for modification, packaging, and secretion of molecules.
– The endoplasmic reticulum is a network of membranous tubes and sacs where lipid synthesis(smooth ER) or protein synthesis (rough ER) occurs.

2 a. Nuclear membrane - separates the genetic material from the cytosol / cytoplasm and only allows small, non-polar molecules to diffuse through, and so is protected from other chemical reactions occurring in the cell.

b. The Golgi is composed of membranous sacs specialized for modification, packaging, and secretion of molecules.

3. Compartmentalization is achieved by enclosing specific enzymes, reactants, and structures within membranes as organelles. The organelles are specialized for performing certain specific tasks, such as cellular respiration This allows enzymes and their reactants to be concentrated in a region where they can operate most effectively. It may also help cell integrity by keeping damaging hydrolytic enzymes isolated.

4. Eukaryotes have a higher level of organization than prokaryotes. Eukaryotic cells can be larger because compartmentalization allows reactions and the molecules required for those reactions to be isolated and regulated rather than dispersed, which is less efficient and would make maintaining the reactions difficult.

87. Techniques in Cellular Visualization (Page 139)

1. When cells are broken apart, the organelles are distributed through the cytoplasm. The ultracentrifuge can spin the liquid at extremely high speeds to generate G force. This is able to separate the same type of organelles enabling collection of one type, allowing for clear visualization.

2. The organelles have different densities depending on their size and type. The particular speed of the ultracentrifuge force created by spin will determine which organelles are separated from the initial homogenate to be later collected.

3. For example; new technology in molecular imaging, such as electron microscopes (scanning SEM or transmission TEM) has allowed scientists to visualize and study the smallest components of the cells, and how they function and communicate with each other - which was not possible with optical microscopes.

88. Adaptations in Mitochondria and Chloroplasts (Page 140)

1. The cristae are formed from the inner mitochondrial membrane, and provide a large surface area, due to their invagination, for the attachment of the proteins and enzymes involved in the electron transport chain component of respiration. This increases the reaction rate, making respiration more efficient.

2. Substances, including enzymes and pyruvate produced in the cytoplasm of the cell, are compartmentalized within the thick liquid-like matrix, so they are concentrated and made available for the Krebs cycle.

3. It separates the matrix from the outer membrane space and allows a proton gradient to be produced across the inner membrane that can then be used for the synthesis of ATP.

4 a. Stroma

b. Stroma lamellae

c. Outer membrane

ISBN: 978-1-99-101423-8

d. Granum

e. Thylakoid

f. Inner membrane

5. Chlorophyll is a membrane-bound pigment found in and around the photosystems embedded in the membranes. The internal membranes provide a large surface area for binding chlorophyll molecules and capturing light. Membranes are stacked in such a way that they do not shade each other.

6. Thylakoids anchor the pigment molecules and enable light energy to be captured in the light dependent reactions. Water bonds are broken (photolysis) producing ATP. Thylakoids contain small 'pockets' of fluid filled with concentrated enzymes and substances. Stroma is a liquid medium in which the light independent reactions (Calvin cycle) occur, using the ATP to provide energy for the reaction that converts substances such as carbon dioxide into glucose.

7. Chemical reactions, like the Calvin cycle, require enzymes and other metabolic substances in sufficient concentrations in order to proceed effectively. These substances are compartmentalized into concentrated fluid, and in a confined space, and made available to the sites of reaction.

89. The Nucleus and Endoplasmic Reticulum (Page 142)

1. Compartmentalization prevents contents of the cytoplasm contacting the contents of the nucleus. It keeps the various reactions in each space separate.

2. The replicated chromosomes need to separate into two sets and move to opposite ends of the cell, prior to it dividing. The nuclear membrane would prevent this movement. It reforms once new cells/gametes are forming.

3. Protein synthesis occurs in both ribosomes, with mRNA moving to them from the nucleus to be translated. Free ribosomes are located in the cytosol only - and synthesize protein that is then used within the cell. Bound mitochondria on endoplasmic reticulum synthesize protein to be transported out of the cell in vesicles.

90. Membranes and the Production of Proteins (Page 143)

1 a. Assemble and process the polypeptides into proteins destined for secretion. Attachment of carbohydrates to the proteins and packaging of the glycoproteins into transport vesicles.

b. A protein that forms around membrane covered vesicles containing protein / glycoprotein for transport - enables the membrane to detach from ER. Breaks down after detaching and reused.

c. These bud off the ER and move substances to the Golgi apparatus, and from the Golgi towards the cell membrane and out of the cell. The membrane forms from the ER membrane.

d. Receives transport vesicles from the ER. Modifies, stores, and transports molecules for export around or from the cell.

91. Stem Cells and Cell Specialization (Page 144)

1 a. Potency- ability to differentiate into other cell types.

b. Self renewal - ability to maintain an unspecialized state.

2 a. The ability to differentiate into any cell in the organism. Found in the zygote in animals and meristems in plants.

b. The ability to differentiate into any cell except extra-embryonic cells e.g. the placenta. Found in the embryo.

c. Ability to differentiate into a limited number of cells related to the tissue of origin (e.g. blood cells). Found in bone marrow, skin, bone.

3. Embryonic stem cells are pluripotent and can form any cells of the body except placental cells. Adult stem cells are multipotent, they have already differentiated from embryonic stem cells into a main 'category' of cell and can divide only into a limited number of specialized cell types, e.g. those of the blood, bone, epithelium.

4. Adult stem cells (ASC) are found as stem cell niches in several types of tissues - in the bone marrow, where they can further differentiate into red/white blood cells and platelets, in the central nervous system, developing into neurons and packing cells (glial), in adipose fat tissue, where cells can develop into bone cells (osteoblasts), fat cells, and cartilage cells (chondrocytes). Stem cells in umbilical cord blood can differentiate into any type of adult cell.

5. Genes for phenotype are coded into each cell, but gene expression determines whether they are turned off or on. Concentration gradients of morphogens influence gene expression, and certain phenotypes leading to structures will be expressed in stem cells (or not) depending on concentration, determining final position.

92. Comparing Human Cell Sizes (Page 146)

1 a. Cardiac: 100 - 150 μm Smooth: 20 - 200 μm

Skeletal: 2.0×10^7 μm - 3.0×10^7 μm

b. Without scales, information about the relative size can be missing. Objects that are many times larger or smaller to each other, i.e. cardiac muscle cells and skeletal fibres, can not be drawn to scale on page.

93. Constraints to Cell Size (Page 147)

1 a. 7.6 cm^2

b. 9.6 cm^2

c. 22.14 cm^2

2 a. 2.0 cm^3

b. 2.0 cm^3

c. 2.0 cm^3

3. Sphere = 7.6/2 = 3.8:1
long cylinder = 9.6/2 = 4.8: 1
Disc = 22.14/2 = 11:1.

All three cells have the same volumes but different SA:V ratios. Changing the shape to thinner (cylinder) or flatter (disk) increases the

ISBN: 978-1-99-101423-8

surface area greatly, even if the volume remains the same. This greatly increases the ability of the cell to obtain nutrients and dispose of wastes because a high surface area allows many molecules to diffuse into and out of the cell.

4. The volume inside the cell reduces as the cell becomes smaller. The surface area also reduces as the cell becomes smaller. Substances need to diffuse 1cm to reach the centre of the smallest cell, but they need to diffuse 2.5cm to reach the centre of the largest cell. SA:V decreases, the larger the cell.

5 a. SA: $6 \times 2^2 = 24\ cm^2$
V: $2^3 = 8\ cm^3$
SA:V 24 : 8 = 3 : 1

b. SA: $6 \times 3^2 = 54\ cm^2$
V: $3^3 = 27\ cm^3$
SA:V 54 : 27 = 2 : 1

c. SA: $6 \times 4^2 = 96\ cm^2$
V: $4^3 = 64\ cm^3$
SA:V 96 : 64 = 1.5 : 1

d. SA: $6 \times 5^2 = 150\ cm^2$
V: $5^3 = 125\ cm^3$
SA:V 150 : 125 = 1.2 : 1

6. The cell can be represented by a cube, and therefore the surface area and volume can more easily be calculated mathematically. The cube shape is close enough to approximate the cell. Using these models, the comparison in SA:V ratios between sizes can also be more easily compared. This might be difficult if cells were all different shapes from each other. The principle can be clearly demonstrated with the models.

94. Investigating the Effect of Cell Size (Page 149)

1. If diffusion is less efficient for cubes with a lower surface area: volume ratio, then larger agar cubes will have a larger ratio of 'no colour change' volume to 'colour change' volume than smaller agar cubes.

2. 1) Cut the phenolphthalein infused agar into three blocks of 1x1x1 cm, 2x2x2 cm & 3x3x3 cm.
2) Fill a beaker with NaOH and place the three agar cubes into the NaOH so that they are completely covered.
3) Let the cubes soak for 5 minutes.
4) Pour off the NaOH and rinse the cubes with water and dry on a paper towel.
5) Cut the 1x1x1 cube in half. Measure the dimensions of the area of no colour change and calculate its volume. Use this value to calculate the volume of the colour change region.
6) Repeat for the 2x2x2 and 3x3x3 cubes.
7) Record and compare the results.

3.

Cube size	Cube volume	Cube surface area	SA:V ratio	Volume of no colour change	Volume of colour change	Ratio of colour change to no colour change
1x1x1	$1cm^3$	$6cm^2$	6:1	$0.125cm^3$	$0.875cm^3$	7:1
2x2x2	$8cm^3$	$24cm^2$	3:1	$3.375cm^3$	$4.625cm^3$	1.4:1
3x3x3	$27cm^3$	$54cm^2$	2:1	$15.625cm^3$	$11.375cm^3$	0.73:1

4. Students will conclude that the ratio of colour change to no colour change decreases as the agar cube volume increases. This indicates a reduction in the efficiency of diffusion into the cell as the cube surface area to volume ratio decreases.

95. Cellular Adaptations to Increase Surface Area (Page 151)

1. Microvilli are multiple folds of the cell's membrane, and can be found on the surface of epithelial cells that line a wide range of structures and organs. These increase the surface area enormously, allowing for extended membrane 'space' for diffusion and transport. Small and large intestine epithelial require them for absorption.

2. The microvilli on the inside epithelial surface has microvilli that increase the transport of water and other substances across the membrane, 'absorbing' them from the filtrate to be returned to the body (via circulatory system). The invaginations of the membrane allow for increased membrane area to form vesicles around the substances, so they can be transported out of the tubule cell.

96. Adaptations of Mammalian Cells (Page 152)

1. Specialization improves efficiency of function within an organism. It saves energy because a specialized cell only needs to maintain the structures and enzymes for a narrow range of function.

2 a. Extended plasma membranes (in microvilli) increase the surface area for nutrient absorption.

b. Electrically excitable so capable of carrying messages (as electrical signals or impulses).

3. For example: Bone tissue has a number of different cells with different functions. The bone matrix / structure is laid down by osteoblast cells, while the mineral deposit is maintained by osteocytes. Reabsorption and remodelling is the function of osteoclasts, where the minerals are recycled. The bone also has osteogenic cells, a type of stem cell, that is found in the bone marrow, and can differentiate into osteoblasts and osteocytes.

4. In cardiac muscle the myofibrils in the shorter individual cells are connected together with desmosomes between cells. In the long muscle fibres, the myofibrils run along the length, and are not connected between 'cells'.

ISBN: 978-1-99-101423-8

5. The branching between cells allows one cell to connect with many more. The cells are able to communicate with each other (through gap junctions) as a network, and coordinate their contractile movement - which is important when the muscles of the heart need to 'beat' together to move blood around.
6. Striated muscle fibres are extremely long (2-3cm), and contain multiple nuclei. This is a result of the 'cells' not dividing from each other during replication, but instead remaining joined to form one long structure, where the myofibrils are continuous. Cardiac cells have one nucleus and undergo cell division in the typical process.

97. Did You Get It? (Page 155)

1 a. Phospholipid
b. Glycoprotein
c. Glycolipid
d. Peripheral protein
e. Integral protein (channel protein)
f. Cholesterol
2. Channel protein - provides a pathway for facilitated diffusion for substances that are unable to diffuse directly through the membrane (due to being polar), or too large to diffuse through the membrane. Often have gates.
3. Cholesterol acts as a packing molecule, interacting with the phospholipids to regulate the consistency of the membrane (preventing both crystallization and excessive fluidity). Seasonal changes - increasing the ratio of unsaturated (bent) lipids to saturated (straight) in warmer weather (and vice versa for cooler weather).
4. Totipotent - found in early zygotes - have the potential to differentiate into any body cell. Pluripotent - found in umbilical cord blood - have the potential to differentiate into any cell, but embrocation cells. Multipotent - found in stem cell niches, such as bone marrow - have the potential to differentiate into a small range of cell types.
5. Compartmentalization in a cell increases the cell's efficiency by grouping reactants and enzymes needed for a specific set of reactions together.
6. Adaptations to increase the surface area of membrane - flatter or longer shape, to increase SA:V ratio (less distance for diffusion), and have microvilli protrusions on the outside of the cell - increase transport.
7. A cage of clathrin forms around a budding section of membrane that contains the substance to be transported (such as glycoproteins). The section is nipped off and frees the resulting vesicle The sections of clathrin then break apart from the vesicle (to be reused) and the vesicle moves towards the outer membrane.
8. Some tissues need to perform more than one function, so different types of cells are differentiated, each performing one of the individual tasks. For example, in alveoli are Type II (secretory) and I pneumocytes (lining).

Theme B: Form and Function

Chapter 7: Organisms

98. Gas Exchange Adaptations (Page 158)

1. The purpose of gas exchange is to meet the metabolic requirements of cells and tissues for respiratory gases by moving oxygen into the body and carbon dioxide out of it.
2. By diffusion across gas exchange surfaces.
3. Molecules are used or removed immediately in the organism or environment maintaining the diffusion gradient.
4. Oxygen is removed from the exchange area, creating a gradient for inward diffusion. Carbon dioxide is constantly brought to the gas exchange area creating a gradient for outward diffusion.
5 a. Provide adequate supply and removal of respiratory gases necessary for an active lifestyle.
b. Larger, more complex organisms cannot rely on diffusion across the body surface - exchange rates would be inadequate.
6 a. Air breathers produce mucus that keeps the gas exchange surface moist.
b. Some water vapour is present in lungs as a result of metabolism.
c. Lungs are internal, preventing drying by the environment.
7. Gills are external structures and need support from a dense medium (water). In air, they would collapse.
8. Ventilation maintains the concentration gradient for the diffusion of respiratory gases via the gas exchange surface.
9 a. Keeping the gas exchange membrane moist. Drying out prevents diffusion of the gases across the membrane.
b. Obtaining enough oxygen, which is in a lower concentration than in the air.

99. Gas Exchange in Fish (Page 160)

1. – Greatly folded surface of gills (high surface area).
– Gills supported and kept apart from each other by the gas exchange medium (water).
– Water flow across the gill surface is opposite to that of the blood flow in the gill capillaries (countercurrent).
– Ventilation of gills maintains diffusion gradients.
2. The gill cover (operculum) acts as a pump, drawing water past the gill filaments.
3. In countercurrent flow, oxygen-rich water flows over the gill filaments in the opposite direction from the blood flow through the gill filaments.
4 a. As blood flows through the gill capillaries (gaining oxygen), it encounters blood of increasing oxygen content so a diffusion gradient for oxygen uptake is maintained across the entire gill surface.
b. In parallel flow, the oxygen concentration in the blood and the water would quickly equalise and

ISBN: 978-1-99-101423-8

diffusion into the blood would stop.

5. The gills contain a dense network of blood vessels that maintain a constant flow of blood. As the RBCs / haemoglobin takes up oxygen, it then moves on so 'new' and oxygen depleted haemoglobin is available. The surface area is maximized. Fish maintain a constant flow of oxygenated water (in a counter current flow) across the gill surfaces.

100. Adaptations for Mammalian Gas Exchange (Page 162)

1 a. Gas exchange occurs across the single celled epithelium of the alveoli, which also needs to be in close contact with capillaries. Greater surface area = more gas exchange.

b. Allows for more capacity for alveoli (where gas exchange occurs), as each alveolus is attached to the end of the smallest bronchioles.

c. The gas exchange occurs across the alveoli to the capillary so, as the greater surface area of the alveoli increases the gas exchange rate, so too does the number of capillaries in contact.

d. The surfactant that coats the alveoli surface reduces the surface tension (and creates a barrier to water being in direct contact with the epithelium) to prevent the alveoli collapsing when exhalation occurs.

2. Thin (Type I pneumocyte cell) layer in close contact with circulating thin walled capillaries. Gas (oxygen and carbon dioxide) has very little distance to diffuse across. Secreted surfactant enables the alveoli sac to remain 'inflated' during both inhalation and exhalation.

1 a. Diffusion is passive transport, and the gas moves from high concentration to low concentration across a concentration gradient. The greater the gradient difference, the greater the diffusion rate.

b. Inhalation introduces fresh oxygen in the air to replace that taken away by the red blood cells. Exhalation removes the carbon dioxide built up in the alveoli sacs - which has diffused across from the plasma / red blood cells.

101. Lung Ventilation (Page 164)

1. Breathing ventilates the lungs, renewing the supply of oxygen while expelling air high in CO_2.

2. Breathing is the result of muscle contraction and relaxation that increases and decreases the volume of the thoracic cavity. The pressure changes cause air to move in and out of lungs.

3 a. External intercostal muscles and diaphragm contract. Lung space increases, air flows into the lungs (inspiration). Inflation is detected, inhalation ends. Expiration occurs through elastic recoil of the ribcage and lung tissue.

b. Muscular contraction is involved in both the inspiration and the expiration

4 a. External intercostals and diaphragm.

b. Internal intercostals and abdominal muscles.

5 a. Internal intercostals and abdominal muscles

b. External intercostals and diaphragm.

6. Antagonistic muscles for inspiration are in opposition to those for expiration. When one set is contracting, the other is relaxed.

102. Measuring Lung Volumes (Page 165)

1 a. 3.15 L

b. 3.75 L

c. Results are as expected. Males are generally, physically larger than females, so their lung capacities are also larger.

2 a.

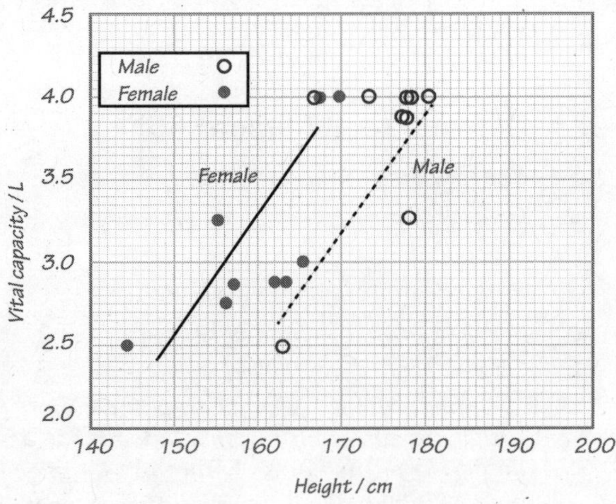

b. There is a positive correlation between height and vital capacity. Taller people generally have larger lung volumes and capacities. Males have larger lung volumes and capacities than females.

Investigation 7.1

Dependent on students:
Typical TV = 400 - 500mL ,
typical IRV = 1900 - 3300mL,
typical ERV = 800 - 1100mL

3. Dependent on students:
Typical IC = 3000mL

4. Dependent on students:
Typical TLC = 6000mL

5. Dependent on students:
Typical VC = 3000 - 5000mL

103. Oxygen Transport and Haemoglobin (Page 167)

1 a. HbF = in embryo from 10 weeks in-utero to completely gone by 6 months old baby. HbA = increasing in proportion to replace HbF from in-utero - and completely by 6 months.

b. Both forms have 4 subunits of protein to make up haemoglobin. HbF = two alpha and two gamma subunits, while HbA = two alpha and two beta subunits.

c. HbF = higher affinity for oxygen - need to rely on oxygen passing through placenta while in-utero. HbA = lower affinity for oxygen, but once born there is an operational respiratory system.

2. HbF = higher affinity for oxygen - need to rely on oxygen passing through placenta while in-utero. HbA = lower affinity for oxygen, but once born there is an operational respiratory system.

3 a. In the alveoli and capillaries leaving the lung.

b. Capillaries leaving tissues, and cells of the tissues.

ISBN: 978-1-99-101423-8

4. *The binding of one oxygen molecule to haemoglobin increases the ease (affinity / attraction) of binding further oxygen molecules to occupy all 4 haem group spaces.*

5 a. *The affinity for the first oxygen molecule in low oxygen tension is lower (less steep gradient in first part of graph) than when oxygen tension rises (steeper gradient) - until HfA is saturated and can accept no more.*

b. *There is only one sub-unit , so myoglobin shows no allostery and therefore no difference in affinity to oxygen - the gradient only slows once myoglobin is saturated and can accept no more oxygen.*

6. *The Bohr effect occurs when actively respiring tissue consumes a lot of oxygen and generates a lot of CO_2. This lowers tissue pH, causing more oxygen to be released from the HfA to where it is required. This effect can be seen as a shift to the right in a dissociation (oxygen un-binds from haemoglobin) curve.*

104. Leaf Adaptations for Gas Exchange (Page 169)

1. *The stomata are openings into the inside tissue of leaves that allows the flow of gases (carbon dioxide and oxygen) in and out for the purposes of respiration and photosynthesis that occur in cells. The open/close of stomata is usually controlled: guard cells use an active transport of ions to become turgid (open stomata) or flaccid (close stomata) to reduce transpiration rate.*

2.

Sectional diagram of plant leaf structure (cross-section)

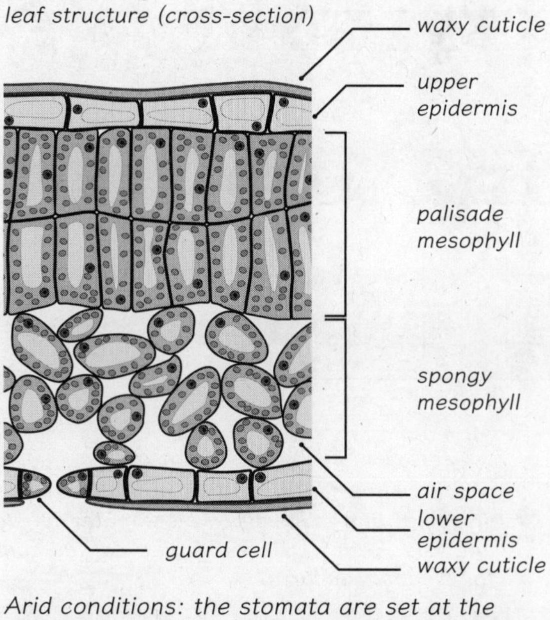

3. *Arid conditions: the stomata are set at the bottom of pits in the stem or leaf = maintains humidity directly around the stomata / reduces airflow -and therefore reduces transpiration when the stomata is open. Aquatic conditions: stomata are found on the top of the leaf and have no guard cells (reduced need for transpiration) and therefore maximum opportunity for gas exchange.*

4 a. *Cactus*

b. *The cellular tissue in the stem contains chloroplasts (with chlorophyll) and stem pores / stomata allow gas exchange with cells.*

5. *The palisade cells are packed most efficiently to allow as many chloroplasts to be exposed to light (for photosynthesis) as possible. The spongy cells are spaced so air can circulate around them and access the cells in the leaf.*

105. Gas Exchange and Transpiration (Page 171)

1. *Diffusion through stomata.*
2. *Cells in the leaf epidermis.*
3. *An alternative entry for gas exchange (to enter plant) if stomata are absent.*
4. *Gas exchange requires an opening into the leaf, usually through a stoma, so that gas (oxygen and carbon dioxide) can diffuse in and out and access all cells (for respiration and photosynthesis). Plants draw in water through their roots (osmosis) but when the stomata are open, water can evaporate and leave the plant (transpiration). Low humidity, wind, and higher temperatures can increase transpiration rate. Plants need to balance the opening of stomata for photosynthesis to occur, with the need to conserve water and reduce transpiration rate - most plants have stomata closed during the night and if they undergo excessive water loss.*

106. Investigating Stomatal Density (Page 172)

1 a. *To examine the relationship between stomatal density and plant leaf adaptations.*

b. *The stomatal density (stomata per cm^2) will be lower in plants adapted to drier environments than in plants adapted to wetter or more moderate environments.*

2.

	Number of stomata per mm^2 lower surface					Number of stomata per mm^2 upper surface				
	Count number					Count number				
Plant name/type	1	2	3	4	Mean	1	2	3	4	Mean
Sunflower	*176*	*180*	*165*	*172*	*172*	*63*	*75*	*80*	*65*	*71*
Germanium	*45*	*46*	*61*	*57*	*52*	*17*	*20*	*22*	*18*	*19*
Garden bean	*276*	*228*	*257*	*238*	*250*	*38*	*50*	*36*	*35*	*40*
Corn	*89*	*113*	*105*	*96*	*101*	*67*	*55*	*59*	*59*	*60*

3 a. *Student answer.*

ISBN: 978-1-99-101423-8

b. *Student answer.*

4 a. *Student answer. There may be no clear relationship.*

b. *There may be factors affecting transpiration rate other than stomatal density. Shading and local humidity around close together leaves may affect transpiration. Also, each plant may have its own physiological processes to regulate stomatal opening and transpiration rate.*

5 a. *On the bottom of the leaf.*

b. *Having most stomata on the bottom on the leaf reduces transpiration while still allowing the entry of CO_2 (the humidity is higher on the leaf underside and the lower leaf surface does not receive direct sunlight).*

6. *Student answer: Students may find that plants from arid areas have fewer stomata than plants from wetter areas. Stomata may be arranged in specific ways based on their lifestyle (e.g. in rows or sunken in pits). Samples that have been taken from aquatic plants may have stomata that are located on the top of the leaf (upper epidermis) and may not have guard cells at all.*

7. *The count of stomata density was taken three or more times, with the mean calculated. This reduces the likelihood that any random area of the leaf that has greater or lower stomatal density than expected will influence the results.*

107. Arteries (Page 174)

1 a. *Tunica externa*

b. *Tunica media*

c. *Endothelium*

d. *Blood (or lumen)*

2. *Thick, elastic walls are required in order to withstand and maintain the high pressure of the blood being pumped from the heart.*

3. *It helps to regulate blood flow and pressure. By contracting or relaxing, it alters the diameter of the artery and adjusts the volume of blood as required.*

4. *Arteries have layers of muscle and elastic tissue which enable them to expand (vasodilation) or constrict (vasoconstriction) to regulate the blood pressure by increasing or decreasing the diameter of the artery lumen.*

108. Veins (Page 175)

1 a. *Veins have less elastic and muscle tissue than arteries.*

b. *Veins have a larger, more expandable lumen than arteries.*

2. *Blood in arteries travels at high pressure, so arteries are thick, strong and stretchy, with a lot of elastic tissue to resist and maintain the pressure. Blood in veins travels at lower pressure so veins do not need to be as strong. They have thinner layers of muscle and elastic tissue and a relatively larger lumen.*

3. *Valves (with muscular movements) help return venous blood to the heart by preventing backflow away from heart.*

4. *Venous blood oozes out in an even flow from a wound because it has lost a lot of pressure after passing through the narrow capillary vessels (with their high resistance to flow). Arterial blood spurts out rapidly because it is being pumped directly from the heart and has not yet entered the capillaries.*

109. Capillaries (Page 176)

1. *Capillaries are blood vessels that exchange oxygen and nutrients in the blood with carbon dioxide and wastes from the cells.*

2. *Capillaries are very small blood vessels forming networks that penetrate all parts of the body. The only tissue present is an endothelium of squamous epithelial cells (only one layer). Capillaries are not under any pressure from blood flow, but must be thin to allow efficient exchange of nutrients and wastes between the blood and tissue. The small diameter causes the resistance to increase and blood flow decreases. The slow blood flow through capillaries allows sufficient time for exchange of gases and nutrients.*

3. *Fenestrations increase the rate of exchange between blood vessels and tissue to ensure sufficient exchange of gases and quick removal of wastes. This enables high metabolic rate to be maintained without the build up of toxic wastes or reduced supply of oxygen and nutrients.*

110. Measuring Changes in Pulse Rate (Page 177)

1. *Student's answer.*

Activity	Resting before exercise.	At 1 minute during exercise	At 2 minutes during exercise	At 3 minutes during exercise	At 4 minutes during exercise	1 minute after exercise	5 minutes after exercise
Pulse Rate	*38*	*64*	*76*	*80*	*88*	*52*	*48*

2. *Although digital tools are likely to be more expensive, they can take heart rate data repetitively while in motion, or even while sleeping, and transmit to a computer for data analysis. Repeated sets of data can be compared for trends over time.*

111. Coronary Occlusions (Page 178)

1 a. *Coronary occlusions are blockages of the coronary arteries.*

b. *The occlusion may partially block the blood flow to the heart, reducing efficiency or causing permanent death to some sections of muscle. Additionally, the plaque may break off and cause a blood clot which can block the artery completely - leading to heart failure and death.*

2. *High blood pressure, high levels of LDL cholesterol,*

ISBN: 978-1-99-101423-8

smoking, and high blood sugar levels.

3 a. *Both men and women show an increase in prevalence as they age, but a much more significant increase in men can be seen. Men, in all age groups, have a higher prevalence than women, especially so in the 75 and over group.*

b. *Women have the highest incidence in the 55-64 age group, which then reduces in incidence rate and stabilizes over the further two age groups. Men see a continued increase of incidence as they age.*

c. *Incidence reduces somewhat in women after the age of 65, however, as expected, the cumulative prevalence increases. This indicates that the disease is likely to be chronic, and not necessarily fatal in the short term. Men see both an increasing incidence and prevalence over time.*

4. *The higher correlation coefficient for age group 39-49 indicates there is a greater correlation between health/lifestyle factors and incidence of CHD than at the older age group (which may be down to other age related issues instead). However, it is not irrefutable evidence that cholesterol and smoking cause of CHD.*

112. The Formation of Tissue Fluid (Page 180)

1. *It bathes the tissues, providing oxygen and nutrients as well as a medium for the transport (away) of wastes, e.g. CO_2. It ensures that these substances can reach each and every cell - which would be unrealistic for capillaries to reach.*

2 a. *Hydrostatic pressure (HP) predominates in causing fluid to move out of the capillaries. Within the fluid is contained oxygen, glucose, ions, and water.*

b. *Increased concentration of solutes and reduction in HP at the venous end lowers the solute potential in the capillary so water and solutes re-enter the capillary. The fluid also contains dissolved carbon dioxide and waste products from the cells.*

3 a. *Returns to the capillaries at venous end*

b. *Enters lymphatic system and eventually drains back into vascular system near the subclavian vein.*

4. *Small molecules, like oxygen, carbon dioxide, ions, and glucose can easily pass from plasma to tissue fluid through the 'leaky' walls of the capillary. These are present in both fluids, although carbon dioxide is removed rapidly and transported back to the lungs. Blood cells and platelets cannot pass through the walls, so are only present in plasma. Some leukocytes are present in both tissue fluid and plasma.*

5. *Around 10% of the tissue fluid does not return to the capillaries, but instead enters the lymphatic system. The fluid enters through gaps in the 'capillaries' of the system - single-celled and overlapping. This fluid then moves into larger pre-collector and then collector ducts. The lymph passes through lymph nodes before returning to the circulatory system.*

113. Single and Double Circulatory Systems (Page 182)

1. *The fish heart is a single pump. Blood flows from the atrium to the ventricle. The mammalian heart is a double pump with two atria and two ventricles separated by a muscular septum. One pump sends blood to the lungs, the other sends blood to the body.*

2 a. *Blood passes from the gills to the body cells through the capillaries, as oxygenated blood.*

b. *Blood flows through the systemic circulation at low pressure, as the passage through the gills increases resistance, and therefore lowers pressure.*

3 a. *Blood returns to the heart, through the pulmonary vein after passing through the lungs.*

b. *Blood flows through the systemic circulation at higher pressure than in the single circuit system, as it enters the heart before being pumped out to the body cells (via the pulmonary artery).*

4. *By passing blood back to the heart from the lungs, the double circuit system is able to maintain high blood pressure. Higher systemic pressures are important for efficient oxygen delivery and processes such as renal filtration.*

5. *Blood flow within vessels can be regulated by the contraction or relaxing of blood vessel walls. This enables animals to restrict blood flow in some areas and increase it in others in response to need.*

114. The Mammalian Heart and the Cardiac Cycle (Page 184)

1 a. *Pulmonary artery*

b. *Vena cava*

c. *Right atrium*

d. *Right ventricle*

e. *Aorta*

f. *Pulmonary vein*

g. *Left atrium*

h. *Left ventricle*

2 a. *QRS complex*

b. *T*

c. *P*

3. *Contraction cannot occur during this period, enforcing a rest so that the heart will not fatigue, nor accumulate lactic acid.*

4 a. *T*

b. *S-R*

c. *T-R*

d. *T-P*

115. Transport of Water Through a Plant (Page 186)

1 a. *The evaporative loss of water from the leaves and stem of a plant.*

b. *Any one of:–Transpiration stream enables plants to absorb the minerals they need (minerals are absorbed with the water and are often in low concentration in the soil). –Transpiration helps cool the plant.*

ISBN: 978-1-99-101423-8

2. *Water loss is regulated by the opening and closing of stomata. Ions move in and out of guard cells by active transport.*

3 a. *The plant would lose water from the cells and wilt. When cells lose too much water, they become flaccid. Lack of water would also result in stomata closing, at expense of respiratory exchange - leading to a reduction in photosynthesis.*

b. *During a prolonged period without water (e.g. a drought), when there was strong wind, or low humidity, increasing transpiration rate.*

4 a. *Transpiration pull: Photosynthesis and evaporative loss of water from leaf surfaces create higher solute concentrations (lower water concentration) in the leaf cells than elsewhere, facilitating movement of water down its concentration gradient towards the site of evaporation (stomata).*

b. *Cohesion-tension: Water molecules cling together and adhere to the xylem, creating an unbroken water column through the plant. The upward pull on the water creates a tension that facilitates movement of water up the plant.*

c. *Root pressure provides a weak push effect for upward water movement.*

5. *Water is moved up the tree by a combination of cohesion-tension, transpiration pull, and root pressure, Together, these processes can move water up to heights of far more than 40 m. Root pressure is the weakest force of the three.*

116. Xylem Tissue and Water Transport (Page 188)

1 a. *The vessel elements and the tracheids.*

b. *Other cells present in xylem tissue include parenchyma (packing and storage cells) and sclerenchyma cells (fibres and sclereids), which provide mechanical support.*

2 a. *The end walls of the vessel elements have perforations that allow water to pass through from one vessel to the next.*

b. *Water passes horizontally between tracheids via thin areas in the wall (pits).*

c. *Vessel elements, because the end-on-end arrangement of the pits allows water to move quickly and unimpeded.*

d. *The thickening provides support and stops collapse of the water conducting vessels so that water conduction is not impeded.*

3. *Xylem forms a continuous tube for the passive transport of water. Because the transport is passive, the cells in xylem do not need to be alive.*

117. Distribution of Plant Tissue (Page 189)

1 a. *Fibre cap*

b. *Phloem*

c. *Xylem*

2 a. *Phloem*

b. *Xylem*

3. *In the roots, the primary xylem forms a star shape in the root centre (with usually 3 or 4 points). The vascular tissue forms a central cylinder through the root (stele). The stele is surrounded by a pericycle. In the stem, the vascular tissues form separate vascular bundles and are arranged in an orderly fashion around the periphery of the stem, with the xylem towards the centre and the phloem towards the outside.*

4. *The cells of the vascular cambium divide to produce the thickening of the stem.*

Investigation 7.4

5. Student's drawing - ensure the students have all the features indicated on page 189.

6. Student's answer - students use the information on the previous page on monocots and dicots to justify answer (whether features are present or absent).

118. Uptake at the Root (Page 191)

1. *Passive absorption of dissolved minerals and water; active transport.*

2. *It allows plants to take up sufficient quantities of minerals from the soil. These are often in low concentration in the soil and a low water uptake would not provide adequate quantities.*

3. *Ions / minerals are actively moved across the root hair membrane - this increases root pressure as the water concentration becomes lower inside the cell - and therefore increases the rate of osmosis as water moves across.*

119. Phloem Tissue (Page 192)

1 a. *Sieve tube elements (sieve tube members).*

b. *Companion cells.*

c. *The companion cell keeps the sieve tube element alive and controls its activity. They are responsible for loading and unloading sugar into the sieve cells.*

2. *Phloem requires active transport, whereas the movement of water and minerals into the xylem is passive.*

3. *Fibres and sclereids provide support and strengthening to the phloem tissue.*

120. Translocation (Page 193)

1 a. *The increase in dissolved sugar in the sieve tube cell increases the solute concentration and therefore its osmotic potential. Water then moves into the sieve tube cells by osmosis, creating a pressure that pushes the sugar solution through the phloem.*

b. *This means the sugar flows from its site of production (leaves) to its site of unloading (roots).*

2. *Food is manufactured in one region of the plant (the leaves) but is required in other regions (e.g. the roots and fruits). It must be transported there.*

121. Movement (Page 194)

1. *Student's own research, answers will vary depending on the species they have chosen.*

ISBN: 978-1-99-101423-8

Adaptations should be clearly identified in each chosen species. Although morphological features will be most obvious, some students may also include behavioural and physiological adaptations. For example, sessile barnacles may have feather-like cirri which beat to draw food in (morphological) and swimming birds have webbed feet for efficient paddling through water to find food (morphological). Birds have physiological adaptations to deliver enough nutrients and oxygen to maintain high energy demands in flight. Some anemones have stinging cells to stop them being eaten (physiological). Some animals hunt at night to avoid the heat of the day or to avoid predators (behavioural).

2. *Evolution by natural selection will cause a predator to improve adaptations for speed for capturing prey. It is likely that the slowest prey will be caught easily. Therefore, faster prey are more likely to survive, and hence pass on their genetic material. Faster predators are more likely to capture prey and survive if food resources are in limited supply and, over time, continue to pass on their adaptations for speed.*
3. *Locomotion requires energy, and larger (land) animals have more weight to move around. Flight is especially energy demanding, and requires a high metabolism, and a constant food supply. However, they can move to where food is, or escape from predators. Sessile animals can survive in environments that do not have a plentiful or regular food supply (i.e filter feeding in marine environments with occasional food in the water).*

122. Swimming Adaptations in Marine Mammals (Page 196)

1. *Breathing out before diving prevents nitrogen from the lungs entering the blood. This prevents decompression sickness (the bends) on ascent. Breathing out before diving also helps to reduce buoyancy, assisting the descent.*
2. *The higher the percentage of blood, as a proportion of body mass, the longer the maximum dive time.*
3. *The blood (with oxygen) moves (down 100%) from body tissue into the central nervous system (cortex and cerebellum up 120%), enabling the seal to continue with effective respiration in CNS cells to retain effective reaction/response (especially if hunting).*
4. *Limbs are adapted into fins (front limbs only and hind limbs reduced or removed for fully aquatic). Tail adapted into flukes (fully aquatic). Body shape becomes longer and streamlined. Body fat increased under skin. Reduced ears.*
5. *Pinnipeds are partially terrestrial - and use their hind limbs for locomotion on ice and land. They also come on to land for birthing. Cetaceans and sirenians are fully aquatic, and hind limbs are a disadvantage to streamlining.*
6. *Fish and their ancestors have always been fully aquatic - therefore, vertical plane fins are most effective with the side-to-side (lateral undulation). Swimming reptiles have evolved from terrestrial ancestors, but the limbs were extended out from the body so they maintained their side-to-side locomotion. Mammals have limbs underneath, with up/down spine flexing - and this has been retained by mammalian marine animals as well.*

123. Exoskeletons and Endoskeletons (Page 198)

1. *Exoskeletons are found on the outside of the organism. They must be shed in order for the organism to grow. Endoskeletons are found inside the body. They are able to grow with the body, and are comprised of living tissue with vascular support.*
2. *The skeleton gives the muscles a rigid attachment point. The muscle is attached to the bone with strong tendons, composed of tough connective tissue. Jointed skeletons allow movement when muscles contract to move the skeleton about the joint. The bone is rigid, and in limbs that require the most movement, the bones are long and straight. The muscle is normally attached to one 'fixed' bone structure, and one movable bone (usually an appendage).*

124. The Sliding Filament Theory (Page 199)

1 i. *C*

ii. *D*

iii. *E*

iv. *A*

v. *B*

2 a. *By changing the frequency of stimulation, so that fibres receive impulses at a greater rate .*

b. *By changing the number and size of motor units recruited (a few motor units = a small contraction, maximum number of motor units = maximum contraction).*

125. Skeletal Muscle Structure and Function (Page 200)

1 a. *The banding pattern results from the overlap pattern of the thick and thin filaments (dark = thick and thin filaments overlapping, light = no overlap).*

b. *The I band = Becomes narrower as more filaments overlap and the area of non-overlap decreases.*

The H zone = Disappears as the overlap becomes maximal (no region of only thick filaments).

The sarcomere = Shortens progressively as the overlap becomes maximal.

2 a. *Relaxed*

b. *The I band can clearly be seen. If the muscle was contracted, it would not be visible.*

3. *Titin is attached to the end of the myosin fibre and secured to the Z line. It has elastic properties. When the myosin / actin is stretched, due to contraction of an antagonistic muscle the titin acts to resist over extension and stabilize the myofibril. The stretching creates a source of potential energy, which is then*

ISBN: 978-1-99-101423-8

released to facilitate the myofibril contracting. This increases the strength of the contraction / reduces energy requirements for the contraction.

126. Antagonistic Muscles (Page 202)

1 a. External intercostal muscles - they lift the ribcage (against gravity) and are active for both passive and active inhalation.

b. Internal intercostal muscles - they are not used in passive breathing, but in active breathing they act as the antagonist muscle that pulls the ribs down and facilitates exhalation.

c. The innermost intercostal muscles are used to support the action of the internal intercostals when active exertion is required.

2. Muscles cannot push, therefore if a body part needs to move both back/forward, up/down etc, then two (or more) sets of antagonistic muscles are required to pull in both directions (one contracts while one relaxes).

3. For inhalation of breath the ribcage needs to expand up and outwards - so the external intercostal muscles contract - this creates pressure with the diaphragm contracting to draw air (with oxygen in). During passive breathing the ribs drop down once the external intercostal muscles relax. During active breathing - air is expelled under force - so the internal intercostal muscles contract and pull ribs down to reduce capacity in the ribcage.

4. They are in oblique orientation to the ribs. The torque created by their different position either lifts the ribs up (and hence outwards) - external, or downwards (and hence inwards) - internal.

5. Titin can build potential energy when stretched (due to the antagonistic muscle stretching), and then release it to contribute to the force of the contraction in the muscle. Titin adds force to the muscle contraction.

127. Movement About Joints (Page 204)

1 a. B

b. D

c. E

d. A

e. C

2. Synovial fluid and cartilage jointly act to facilitate free joint movement. Synovial fluid lubricates the joint and reduces friction between the articular cartilage of the bone ends during movement. It also absorbs shocks, and provides nutrients to and removes wastes from the cells of the cartilage. The cartilage cushions the bone ends and, because some synovial fluid is stored within the cartilage, it also provides a store of lubrication.

3. Bones join together at joints providing stability, strength, and flexibility of motion. The hip joint is a ball and socket joint and allows free movement around a large number of axes with one common centre. Tough, connective tissue structures called tendons, join muscle to bone. Muscles are contractile tissue that pulls against bones to enable movement of the skeleton. Strength and stability are given to bones with additional ligament connections.

4. Students should explain that muscle attachment from the neck to the arm allows a full 180 degree movement in shoulder flexion, whereas shorter muscles, paired with ligaments holding the humerus bone in place and clavicle, reduce the angle when shoulder abduction occurs.

128. Did You Get It? (Page 207)

1. In both systems, oxygen needs to enter the capillaries by diffusion, and carbon dioxide needs to be removed from capillaries. This occurs across a concentration gradient. In fish, the oxygen is dissolved in water, and the large surface area of the gills, plus active movement of the water across the surface, allows oxygen to diffuse across the gill into blood (in capillaries) flowing in counter current. In mammals the oxygen is carried in the air through to the alveoli. Oxygen diffuses across cell layers into blood.

2. Single layer, thin Type I pneumocyte cells - small distance for diffusion. Type II pnuemocyte cells secrete surfactant to prevent alveoli collapse.

3. Capillaries. Thin, single-celled walls that allow easy diffusion across them, but still large enough to allow flow of red blood cells (with haemoglobin) to be transported through them. The small diameter allows them to be situated close in a network around the alveoli in order for gas exchange to occur.

4. The Bohr effect occurs when actively respiring tissue consumes a lot of oxygen and generates a lot of CO_2. This lowers tissue pH, causing more oxygen to be released from the HfA to where it is required. This effect can be seen as a shift to the right in a dissociation (oxygen unbinds from haemoglobin) curve.

5. Stomata must be open (controlled by guard cells) in order for gas to diffuse in and out of the leaf. However, when stomata are open this also causes transpiration, as water is evaporated out of the leaf. The plant must balance gas exchange requirements with the need to limit transpiration when water is in limited supply.

6. Synovial joint. (Ball and socket - shoulder). The bone ends that connect are lined with cartilage to reduce wear and tear - with fluid filled spaces to reduce friction when moving.

7. Muscles move in pairs, but in different directions - this is because muscles can pull, but not push. When one muscle set contracts, the other set must relax. The muscles contract alternatively to generate repetitive movement.

Theme B: Form and Function

Chapter 8: Ecosystems

129. Habitat (Page 209)

1. A habitat is the area in which an organism lives, including all the biotic and abiotic factors. An ecosystem may include all the biotic and abiotic factors but also the interactions between them.

ISBN: 978-1-99-101423-8

An ecosystem may include many habitats.

2. Habitats vary in scale, depending on the space needed for the organism. For some oceanic fish that need to swim long distances to find prey, their habitat is vast and relatively homogeneous; for microbes, the habitat could be a specialized part of a mammal's gut.

3. The habitat describes what is in the area. A city habitat will have people, concrete buildings, streets, vehicles, air quality factors, temperature variations etc. A city ecosystem includes all the interactions between the people, the buildings they enter and leave, the refuse they generate etc.

130. Plant Adaptations (Page 210)

1. One that aids in the conservation of water or survival and reproduction in arid regions.

2. Any 3 of: Modification of leaves to reduce losses via transpiration (e.g. spines, curling, leaf hairs). Shallow, extensive fibrous root system extends area for absorbing water and collects condensation.

   Water storage in stems or leaves. Rounded, squat shape reduces surface area for water loss.

3 a. It reduces the diffusion gradient between the leaf and the air, so there is less tendency for water to leave the plant.

b. Trichomes create a layer of still air at the leaf surface and this reduces the gradient for diffusion of water from the leaf.

4. It means that, relative to the plant's volume, there is very little surface area over which water can be lost.

5 a. Cable roots create a firm platform. (A wide root base is more secure, especially in soft mud).

b. Prop roots which act as buttresses for extra support.

6. Pneumatophores, which extend above the surface of the mud, allow entry of oxygen from the air.

7. A high risk of dehydration because the osmotic gradient is from the plant cells to the saline environment.

8. Any of: Storing the salt in older leaves which are then shed from the plant. Waxy suberin on the root cells allows water, but not salt, to enter, so most salt is excluded. Storing salt in the cell vacuoles and maintain high cell solute levels to reverse the osmotic gradient. Secreting salt through glands in the leaf surface.

131. Tolerance and Population Distribution (Page 212)

1. An organism will occupy a habitat according to its tolerance range for a particular suite of conditions. Organisms will be found where all or most of their requirements are met and will avoid those regions where they are not. Sometimes, a single factor, e.g. pH, will determine presence or absence.

2 a. Dissolved oxygen

b. During the autumn/summer (Dec - May) months, the zooplankton stay near the surface of the lake. During the winter (July-Sep), the zooplankton are spread throughout the lake.

3. Water (lack of), temperature (too high), light levels (likely too high or direct).

132. Sampling and Sensors (Page 213)

1. Dataloggers are able to collect and store a range of data at set intervals automatically and can be left in the field. The data can be downloaded at a later date.

2. Abiotic factors need to be sampled so that an idea of the habitat that the organisms being studied live in can be built up. This will allow analysis of the factors that affect an organism's distribution.

3. Thermometer (water and air temperature), salinity meter, light meter, dissolved oxygen meter (for dissolved oxygen in rock pools).

133. Transect Sampling (Page 214)

1 a. With belt transects of 10 m or more, sampling and analysis using this method is very time consuming and labour intensive.

b. Line transects may not be representative of the community. There may be species which are present but which do not touch the line and are not recorded.

c. Belt transects use a wider strip along the study area so there is less chance that a species will not be recorded.

d. Transect sampling is not a suitable technique when the species of interest are highly mobile.

2. Decrease the sampling interval. If no more species are detected and the trends along the transect remain the same, then the sampling interval was adequate.

3.

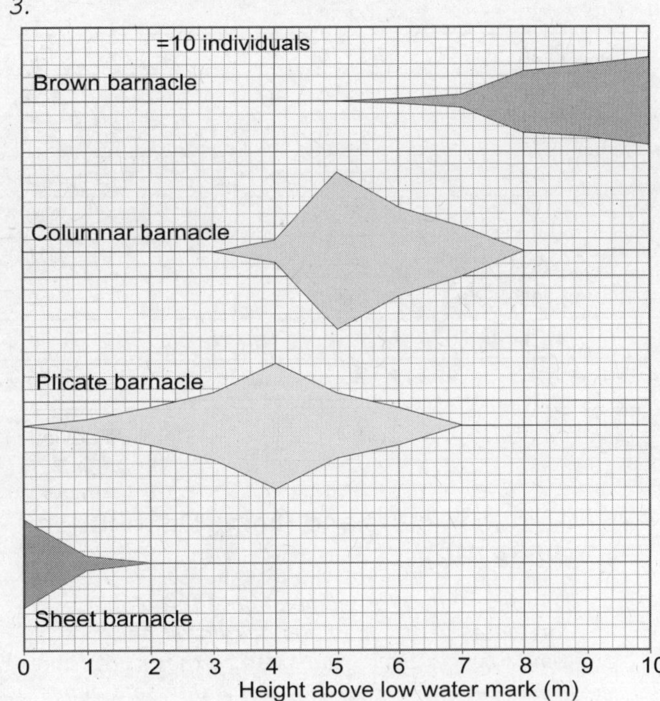

ISBN: 978-1-99-101423-8

134. Abiotic Factors and Population Distribution (Page 216)

1 a. *Increases from LWM to HWM*

b. *Increases from LWM to HWM*

c. *Decreases from LWM to HWM*

d. *Increases LWM to HWM*

2. *Rock pools may have very high salinity due to evaporation after long exposure times without rain.*

3 a. *B will receive the full force of waves moving inshore, A will receive only milder backwash, C will experience some surge but no direct wave impacts.*

b. *A and B will experience greater temperature variations depending on tides, time of day, water temperature. C is more protected and will not experience direct sunlight.*

4. *See table below*

5. *An r of 0.60 tells us that there is a reasonably strong positive correlation between the temperature of the pond water and the number of Hexarthra present.*

4.

***Hexarthra* no. (x) / per L**	**Temperature (y) / °C**	**$(x - \bar{x})^2$**	**$(y - \bar{y})^2$**	**xy**
36.21	19.75	*7,933.72*	*4.90*	*715.15*
33.76	17.53	*8,437.82*	*0.00*	*591.81*
10.83	15.05	*13,176.19*	*6.18*	*162.99*
1.88	14.40	*15,310.99*	*9.83*	*27.07*
0.33	11.73	*15,696.98*	*33.71*	*3.87*
2.40	11.05	*15,182.58*	*42.07*	*26.52*
0.35	9.23	*15,691.97*	*68.99*	*3.23*
0.08	8.75	*15,759.69*	*77.19*	*0.70*
0.00	12.35	*15,779.78*	*26.89*	*0.00*
0.04	13.13	*15,769.73*	*19.41*	*0.53*
0.00	14.15	*15,779.78*	*11.46*	*0.00*
0.21	14.63	*15,727.07*	*8.44*	*3.07*
0.29	15.98	*15,707.01*	*2.42*	*4.63*
5.72	19.63	*14,375.43*	*4.38*	*112.28*
4.39	18.00	*14,696.13*	*0.22*	*79.02*
7.42	19.80	*13,970.67*	*5.13*	*146.92*
72.87	23.33	*2,782.31*	*33.57*	*1,700.06*
443.38	23.30	*100,972.94*	*33.22*	*10,330.75*
34.38	22.30	*8,324.30*	*22.70*	*766.67*
147.58	25.88	*482.35*	*69.62*	*3,819.37*
947.64	24.58	*675,720.83*	*49.62*	*23,292.99*
573.47	22.90	*200,571.77*	*28.77*	*13,132.46*
444.63	20.95	*101,768.91*	*11.66*	*9,315.00*
338.25	21.10	*45,212.54*	*12.70*	*7,137.08*
34.33	18.90	*8,333.43*	*1.86*	*648.84*
$\bar{x}$=125.62	*$\bar{y}$=17.54*	*$\Sigma (x - \bar{x})^2$ = 1,373,224.92*	*$\Sigma (y - \bar{y})2$ = 584.96*	*Σ xy = 72,021.02*

Standard deviation x = *234.4*	Standard deviation y = *4.8*	*r = 0.60*

135. Abiotic Factors and Biome Distribution (Page 219)

1. *The distribution of biomes about the globe correlates with conditions of rainfall and temperature. Similar biomes are therefore found in similar conditions, e.g. tropical desert biomes are found in areas of high temperature and low rainfall, while tropical rainforests are found in areas of high temperatures and high rainfall.*

2. *Because rainfall and temperature are not even. Mountains cause rain on one side and rain shadows on the other and affect air temperature. Large bodies of water can also affect temperature and rainfall.*

3. *By affecting the air travelling over it, either adding or removing water and raising or lowering the temperature. High mountains deflect wind and remove water from the air, creating dry air and elevated temperatures on their leeward slopes. Bodies of water tend to remain at the*

ISBN: 978-1-99-101423-8

same temperature and so modify nearby land temperatures, keeping them more equable.

4. The angle the surface of the Earth presents to the Sun changes from almost 90° at low latitudes (the equator) to almost 0° at higher latitudes (the poles). This causes the Sun's energy to be spread out over a greater surface area at the poles, so that a km^2 of land at the poles receives much less solar energy than a km^2 of land at the tropics.

136. The World's Terrestrial Biomes (Page 220)

1. These are rainshadow areas, in the leeward side of mountains, in the path of rain-bearing winds. Much of the precipitation is dropped at altitude in the mountains, so there is little precipitation in adjacent lowland areas.
2. Water distribution is affected by solar heating, winds, and the geography of the land – especially mountains. Consistent rains are limited by wind patterns to the tropics and the temperate zones and along the windward sides of mountains. Plant material is largely limited to where there is reliable rain and so the most dense plant populations (and hence animal populations) are limited to where the rain is.

137. Marine Biomes (Page 222)

1. Around the equator in areas of shallow water (edges of islands, and shallows seas)
2. The amount and quality of light would not be good enough for photosynthesis and for the coral to form.
3. Warmer sea water will be outside the tolerance of the corals. Falling pH makes it difficult for organisms that use calcium carbonate to build shells.
4. These waters are both warm and have enough light for the zooxanthellae to photosynthesise.

138. Adaptations to Tropical Environments (Page 223)

1. Student's own research. Answers should address the adaptations organisms have evolved in response to the tropical rainforest environment, e.g. light levels, temperature, humidity, density of foliage etc. Adaptations may include ways to cope with high rainfall, low soil nutrients (plants), low light levels (for photosynthesis or visibility during hunting), and competition with other species. Adaptations could include physiological, morphological, and behavioural adaptations.

139. Adaptations to Desert Environments (Page 224)

1. Water
2. Adaptations include: Highly efficient kidneys able to conserve water, large ears to radiate heat, being nocturnal or crepuscular (active at dawn and evening). Storage of fat in humps or other parts of the body. Fur shields parts of the body exposed to sunlight, but is thin in parts that are in shade.
3. Plants might have: deep roots to reach water, extensive roots near the surface to absorb water from rain, reduced leaves, or spines, photosynthetic stems instead of leaves, hairs or thick wax covering leaves to reduce transpiration and reflect light.

140. The Ecological Niche (Page 225)

1 a. The realized niche of a species can be narrower or broader, depending on the constraints that other species place on physical space and resource use.

b. Competition will exert the greatest constraint on the extent of an organism's niche. To a lesser extent, so too will parasitism, predation, and disease.

2. Interspecific competition will result in a divergence in resource use curves and narrower niches. Intraspecific competition will broaden niches, forcing individuals to occupy the whole extent of their tolerance range.

141. Dealing With Different Levels of Oxygen (Page 226)

1 a. Obligate anaerobe

b. Facultative anaerobe

c. Obligate aerobes

2. A facultative anaerobe will be able to take advantage of both aerobic and anaerobic conditions, depending on which conditions they find themselves in, and which are the most efficient.

142. Modes of Nutrition (Page 227)

1 a i. Free energy in sunlight.

a ii. Carbon dioxide

b i. Inorganic compounds, e.g. elemental hydrogen or sulfur.

b ii. Carbon dioxide.

c. i. Organic carbon (usually glucose).

c ii. Organic carbon (usually glucose).

2. It would give the organism the advantage of being able to draw energy from light or chemicals depending on the environment (e.g. in low light, energy can be derived chemically).
3. Holozoic feeders ingest food and digest it internally. Saprotrophs obtain their nutrition by the extracellular digestion of dead organic matter.

143. Dentition and Diet in Hominidae (Page 228)

1 a. Biting, clipping

b. Biting and holding (in carnivores). Threat display in hominids (except humans).

c. Chewing, grinding (for mostly coarse plant material)(modified in carnivores for shearing).

2. Teeth are adapted to specific diets and so their shape indicates what they are used for. For example, thin, chisel-like teeth are used for clipping and biting. Flat teeth with ridges are

ISBN: 978-1-99-101423-8

likely used for chewing and grinding up coarse material. In hominids, the molars vary in size between species, depending on diet. Gorillas have very large molars for chewing vegetation. Humans have much smaller molars for chewing a range of soft foods.

3. *Paranthropus robustus has the most coarse diet, based on the skull, jaw and teeth. Homo erectus has the softest diet. Its teeth and jaw are relatively small.*

144. Herbivores and Resisting Herbivory (Page 230)

1. *Piercing mouthparts, chewing mouthparts.*
2. *Ruminants house microbes in a special part of the stomach (the rumen). These are able to break down the cellulose. The ruminant is then able to further digest the material produced (by further chewing and digesting).*
3. *Plants: spine and thorns to deter browsing animals. Animals, mobile tongue to reach between spines, leathery lips to prevent harm from spines.*
   Plants: Production of toxins in leaves to deter chewing animals. Animals: Production of proteins, enzymes to inhibit toxins on leaves.

145. Predator-Prey Strategies (Page 231)

1. *Warning behaviour and spraying chemicals (skunk). Standing guard.*
2. *Stealth (snakes, spiders). Tool use (chimpanzees, ravens), building traps (spiders). Also cooperative (group) hunting in pelicans, lions.*
3. *Mimic benefits because predators universally avoid attacking animals with the same warning colouration, whether they are poisonous or not.*
4. *Antlers, horns, claws, hooves.*
5. *Skunk spray, toxins in wings (butterflies), reactive chemicals (bombardier beetle).*
6. *Heat detection by snakes allows them to detect prey even at night and in the pitch black of burrows.*

146. Plant Adaptations for Gathering Light (Page 232)

1. *The strangler fig germinates in the upper branches of trees. There, it puts out leaves and so gathers light without requiring the growth of a trunk. It puts out roots that envelop the host tree and reach the ground. These roots eventually form a large trunk.*
2. *Forest floor plants often have large leaves to gather the little light available. The leaves are thin, with a single layer of cells containing chloroplasts. Cells beneath this layer also have chloroplasts to capture light passing through the upper layer.*

147. Fundamental and Realized Niches (Page 233)

1 a. *A represents the region actually occupied by Chthamalus, i.e. its realized niche (the less extensive area where Chthamalus is normally).*

b. *When Semibalanus is removed from the lower shore, the range of Chthamalus extends into areas previously occupied by the Semibalanus; Semibalanus normally excludes Chthamalus from the lower shore.*

2 a. *The Semibalanus larvae die from desiccation at low tide when they settle high on the shore.*

3. *A higher sea level will allow Semibalanus to settle higher on the shore, within the current Chthamalus zone. As Semibalanus currently excludes Chthamalus from the lower shore, it is reasonable to assume it will outcompete them on the upper shore if the physical conditions favour its successful settlement there.*

148. Competition and Competitive Exclusion (Page 234)

1. *The competitive exclusion principle is the principle that two species that compete for exactly the same resources cannot coexist and that one (the more able competitor) will always exclude the other (the less able competitor).*

2 a. *To grow rapidly to carrying capacity and occupy the entire volume of culture (nutrient medium with bacterial food).*

b. *P. caudatum occupies the uppermost, well oxygenated region of the culture, whereas P. bursaria occupied the lower, oxygen depleted region.*

c. *Support : each species grew to carrying capacity when cultured alone, but in competition, P. caudatum became extinct. Against: the experiments with P. caudatum and P. bursaria show that two species, apparently with the same niche can coexist. This seems to contradict the competitive exclusion principle.*

149. Did You Get It? (Page 235)

1 a. *0.4 m*

b. *Lichens (various species).*

c. *Most mosses need higher moisture and lower light and temperatures than lichens.*

2. *The tooth shows a flat surface with ridges. This indicates it is for chewing and grinding coarse material. It is likely a molar. From the photo, it can be seen the tooth is very large. The animal will require a large jaw and skull to carry the tooth. The body will need to be very large to carry the skull and accommodate the muscles to hold and move it. The animal is likely to be slow moving (herbivore and large). (It is a mammoth tooth).*

3 a. *The natural environment where an organism lives including the biotic and abiotic factors.*

b. *Physical factors, including weather, rocks, soil, temperature, etc.*

c. *A heritable structural, behavioural, or physiological feature or trait of an organisms that helps it survive in its environment.*

4. *Poisonous species benefit by having bright colours as predators learn that these colours mean prey is unpalatable. This means poisonous animals are less likely to be attacked in the first place.*

ISBN: 978-1-99-101423-8

150. Summary Assessment (Page 236)

1. *d*
2. *c*
3. *c*
4. *a*
5. *d*
6. *b*
7. *Organisms that are tolerant to a wide range of environmental factors can occupy a wider range of habitats, so they are often widespread in their distribution.*
8. *Phloem in plants is under pressure. When an insect (e.g. aphid) pierces the phloem, the pressure of the sap causes it to flow into the aphid's mouth parts. The mouthparts regulate the pressure of the sap as the aphids feeds so they don't actually suck sap. (Xylem is not under pressure and so sucking is required if xylem is pierced).*

9 a. *The curve shows the saturation level of haemoglobin with oxygen as the pressure of oxygen increases. Saturation happens rapidly from 2 - 8 PO_2*

b. *The curve only reaches a saturation level of about 45%. This means that, even at high O_2 levels, a person's blood is not able to carry much O_2 due to the low level of functional haemoglobin.*

10 a. *Transpiration pull, water moves up the xylem from the pull of a water gradient caused by water molecules leaving the leaves and leaving spaces into which water molecules move. Cohesion-tension, water molecules cling to each other and move up the xylem by the pull of other water molecules moving. Root pressure, water moving into the xylem in the roots pushes water already in the xylem further up.*

b. *Still air, light shade, room temperature.*

c. *High wind, high light, high temperature, low humidity. All increase evaporation rate from leaves.*

d. *Humid conditions reduce evaporative loss, dark conditions stop photosynthetic production of sugars reducing leaf solute concentration. Both reduce transpiration rate by reducing the concentration gradient for water movement.*

11. *A saprotrophic organism feeds on decaying material. Because a saprotroph feeds on material other than that it has made (it cannot make its own food), it is therefore also a heterotroph.*

Theme C: Interaction and Interdependence

Chapter 9: Molecules

151. The Properties of Enzymes (Page 241)

1. *Metabolism is the sum of all cellular reactions. Enzymes play a major role by bringing reactants (substrates) together, orienting them for more efficient reactions. They also lower the activation energy needed to start reactions.*
2. *Enzymes have an active site that will only allow very specific molecules (sometimes only one) to bond to them. Metabolic reactions can consist of a pathway of many reactions, each step starting with its own reactants and therefore require specific enzymes. Enzyme specificity can be defined as, 'one enzyme is specific to one reaction'. Due to the extremely large number of different reactions in humans (and other organisms), it follows that there is a correspondingly large number of different enzymes.*
3. *Anabolic reactions are 'building' and start with more than one substrate and end with the final product being a combined larger product (macromolecule). Examples include the formation of glycogen, a polymer formed from condensation reactions that join together monomer glucose units. Catabolic reactions are 'breaking' and end with more (and smaller) products than reactants. Digestion is considered catabolic because large fat, protein, and carbohydrate macromolecules are broken into smaller substrates that can enter the bloodstream to be further processed.*
4. *Hydrolysis reactions require a water molecule to be added to a polymer (or macromolecule) - OH on one and H on the other - in bonding the OH/H the bonds between monomers are broken - hence a catabolic reaction. In condensation reactions a water molecule is removed from a substance (OH from one and H from the other which combine into water), and once H_2O is removed, those sections of the substance bond - hence anabolic.*
5. *The enzyme changes shape when a substrate molecule is drawn into the active site. The change in enzyme shape also alters the shape of the active site, allowing an enzyme-substrate complex to form. Once the enzyme has reacted with the substrate, the product is released and the enzyme resumes its original shape.*

6 a. *The active site is the region where substrate is drawn in and positioned in such a way as to promote the reaction. The active site is formed by the protein's tertiary structure which creates a precise configuration in which the amino acid side chains can interact with the substrate.*

b. *The main substrate, RuBP, is the same in each reaction - so even though the reactions are different, the same enzyme can be used in both as the substrate will fit into the active site.*

7 a. *Pepsin would have some activity if the pH of the small intestine remained under 6.5. However, when the pH increased above 6.5 the pepsin would become inactive and it would no longer digest protein.*

b i. *Pepsin would probably become active because it is not denatured at pH 7.0 (only inactive), at pH 1.5 it can reform its active three-dimensional structure.*

b ii. *Pepsin would not regain its activity. At pH 8.5, the enzyme has been fully denatured and lost its three dimensional functional structure. It cannot 'renature'.*

8 a. *Enzyme activity has completely stopped (the enzyme is denatured).*

b. *If the temperature remained too high, the activity rate of enzyme X would stop altogether (enzyme denaturation). The metabolic pathway would not proceed and the individual may die.*

ISBN: 978-1-99-101423-8

152. Factors Affecting Enzyme Reaction Rate (Page 245)

1. *Adding energy destabilizes the bonds in the reactants. This causes them to break and then form new, more stable bonds. (Note: Some products are at a higher energy level than reactants - endothermic).*
2. *Enzymes lower the activation energy for a reaction by influencing bond stability in the reactants and creating an unstable transition state in the substrate from which the reaction proceeds readily. Another energy pathway at a lower level is available (reactions can occur at a lower temperature).*
3. *Energy is required to break the bonds in the reactants between the carbon, oxygen, and hydrogen in glucose and oxygen-oxygen (in reality, not every bond is broken at once, but in steps). Energy is released when new bonds form (between carbon-oxygen and hydrogen-oxygen) this energy is a greater amount than required to initially break the bonds - therefore a NET decrease in energy to make respiration exothermic.*
4. *1.40 cm^3/min*

5 a. *Reactants must be constantly being added to the mix to provide the 'raw materials' for the reaction. Without a limiting factor, the reaction continues at the same rate.*

b. *The reactants are being used up, and not replaced. Concentration of reactants becomes the limiting factor.*

6. *Initially, there is plenty of substrate but as substrate is consumed, the reaction rate falls off. There are fewer reactants, so therefore fewer collisions between substrates - resulting in a decreased reaction rate.*

7 a. *If the substrate is not limiting, the reaction rate will increase as the concentration of enzyme is increased.*

b. *A cell may increase the rate of protein synthesis (transcription and translation) to increase the amount of enzyme present, or inactivate enzymes (e.g. by feedback inhibition) to reduce their activity.*

8. *The reaction rate reduces over time because, after a certain concentration of substrate, the enzymes are saturated by the substrate and the reaction rate cannot increase further.*

9 a. *An optimum temperature for an enzyme is the temperature where enzyme activity is maximal.*

b. *25°C is well below the enzyme's optimum and its activity would be very low. Once the temperature was increased to optimum, the enzyme's activity would increase to maximum, because enzyme structure is not generally irreversibly altered by moderately low temperatures outside the optimum.*

c. *Reaction rate at T°C = 2.4 ÷ 30 = 0.08 cm^3/s*
Reaction rate at T + 10°C = 4 ÷ 27 = 0.15 cm^3/s
Q10 = 1.88

10 a. *Pepsin = 1-2*
Trypsin = approx. 7.5-8.2
Urease = approx. 6.5-7.0

b. *The stomach is an acidic environment which is the optimum pH for pepsin, whereas trypsin works in the alkaline environment of the small intestine. The optimal pH of urease suits a neutral environment (it is found in soil and bacteria and fungi).*

153. Investigating Peroxidase Activity (Page 248)

Investigation 9.1

5.

	Colour reference number						
	0 min	1 min	2 min	3 min	4 min	5 min	6 min
pH 3	0	2	2	3	3	3	3
pH 5	0	2	4	5	6	6	6
pH 6	0	3	3	3	3	3	3
pH 7	0	3	4	4	4	4	4
pH 8	0	3	3	3	3	3	3
pH 10	0	0	0	0	0	0	0

1. *The colour palette can be as a reference for enzyme activity.*
The darker the colour, the greater the enzyme activity.

2.

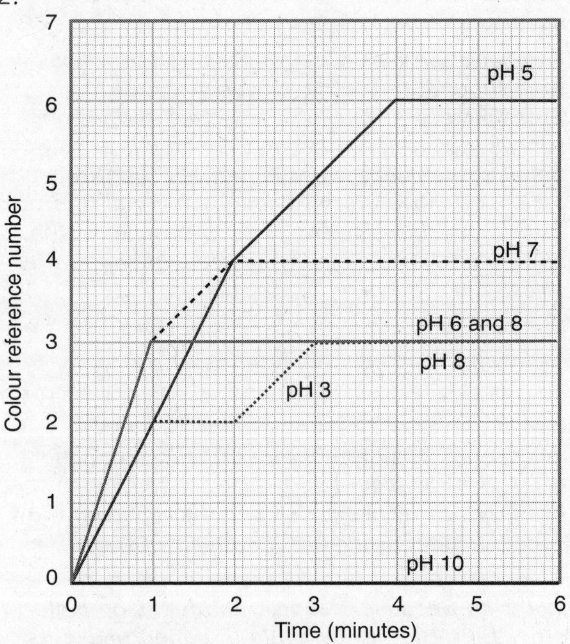

3 a. *Peroxidase has an optimum pH of 5.*
Below and above this pH, the activity of the enzyme is reduced.

b. *There was probably no colour change. At pH 10 the enzyme has probably denatured (lost its 3D structure). This would alter the shape of its active site and substrate would not bind (so reduced or no enzyme activity).*

4. *Students could have used a colorimeter and measured the light absorbance.*
5. *The enzyme begins to catalyse the reaction immediately. If timing does not begin immediately, the recorded time for the colour change will be less than the actual time.*
6. *The peroxidase is written over the arrow because it is a catalyst and it is not used up or changed in the reaction.*

ISBN: 978-1-99-101423-8

7. 1) Prepare substrate tubes by adding 7 mL of distilled water, 0.3 mL of 0.1% H_2O_2 solution, and 0.2 mL of prepared guaiacol solution into each of 6 clean test tubes. Cover the tubes with parafilm and mix.

2) Prepare enzyme tubes by adding 6.0 mL of prepared buffered pH 5 solution and 1.5 mL of prepared turnip peroxidase to each of 6 clean test tubes. Cover the tubes with parafilm and mix.

3) Add 1.6 µL of a 0.1 M lead nitrate solution to 1000 mL distilled water to make a 20 ppb stock solution. Dilute the stock solution of 20 ppb to give 10, 5, 2.5, and 1.25 ppb solutions.

4) Add 1 mL of each lead nitrate solution (20, 10, 5, 2.5, and 1.25 ppb) to each of 5 of the substrate tubes. Tube #6 is the control (no lead nitrate).

5) Combine the contents of substrate and enzyme tubes in 20 mL test tubes, cover with parafilm and place in a test tube rack at room temperature. Start timer and time reaction for 5 minutes. Record colour changes. If working alone or in pairs, you will need to mix and record for each test tube in turn.

6) Compare colours to the reference palette and record the colour reference number. Alternatively, use a colorimeter to measure the absorbance of each solution and provide a quantitative value.

154. Intracellular and Extracellular Enzyme Reactions (Page 250)

1 a. All enzymes are formed in cells - as part of the protein synthesis process. Intracellular enzymes are retained inside the cells in which they are produced in order to facilitate a cellular process. Extracellular enzymes are transported out of the cell (often inactivated) and delivered to the area of the body that they are required.

b. Most metabolism occurs inside cells - as opposed to extracellular reactions that occur in the lumen (spaces) of organs.

2. Intracellular enzymes - as respiration occurs within mitochondria inside cells. There are many different enzymes involved as there are also many different reactions (with different reactant substrates) along the metabolic pathway.

155. Generating Heat (Page 251)

1. Reactions inside the body / cell need to occur within a reasonably stable temperature that is warm enough to provide heat energy for the reaction rate to be sufficient (even with enzymes). Metabolism will be 'slowed' if the temperature is not warm enough (this can be seen in organisms like reptiles that need environmental heat to supplement metabolic heat).

2. Energy passes from substrate to substrate during chemical reactions (chemical potential energy) - but not 100% - as some will be transformed into heat energy and lost to the surroundings.

3. Animals with high energy demands - such as birds and mammals that are very mobile - require a high metabolism to supply the energy needed (via ATP) to their muscles. To maintain the high metabolism, there needs to be a constantly maintained body temperature (thermal homoeostasis), and therefore sufficient heat generation.

156. Metabolic Pathways (Page 252)

1. A metabolic pathway is a series of linked biochemical reactions that carry out processes essential to maintain life.

2. Enzymes control each step of the metabolic pathway.

3. There is a starting substrate (Glucose) and a final substrate (pyruvate) - even though two separate branches form the pyruvate. The final substrate does not enter the metabolic pathway again.

4. ATP is the energy carrier in cells. ATP provides the energy (via a high energy phosphate bond) required for metabolic reactions. ATP is hydrolysed to produce ADP and a free phosphate, releasing energy in the process.

5. The outputs of photosynthesis (oxygen and glucose) are the inputs for cellular respiration. The outputs of cellular respiration (carbon dioxide and water) are the inputs for photosynthesis.

6.

	Linear pathways	Cyclic pathways
Named examples	Glycolysis, coagulation cascade in blood clotting	Krebs cycle (respiration) Calvin cycle (photosynthesis)
Relationship of beginning reactant and ultimate product	Beginning reactant and final product different	Beginning reactant and final product same
Complexity	Relatively simple - one reaction follows another	more complex - often with other metabolic pathways branching off from intermediates

157. Enzyme Inhibitors (Page 254)

1. In competitive inhibition, the inhibitor competes with the substrate for the enzyme's active site and, once in place, prevents substrate binding. In non-competitive inhibition, the inhibitor does not occupy the active site but binds to some other part of the enzyme, making it less able to perform its catalytic function.

2. They are specific to that particular reaction of producing LDL cholesterol - so will not interfere with other reactions, even at high doses. The inhibitor is competitive, so it is reversible - and the cholesterol will be able to be produced once more when suitable levels are reached.

3. Allosteric regulators bind to an allosteric site (not the active site) on an enzyme, changing the shape of the active site. Depending on the type of regulator that binds, the enzyme may be activated or inactivated.

ISBN: 978-1-99-101423-8

4. *Heavy metals are toxic because they permanently stop enzyme activity, rendering the enzyme non-functional. Because they are lost exceedingly slowly from the body, anything other than a low level of these metals is toxic.*

5 a. *Some antibiotics, such as penicillin, are irreversible inhibitors to the enzymes essential for wall synthesis in bacteria. Susceptible bacteria are unable to build cell walls so cannot complete their cell division (growth stops).*

b. *Human cells do not have a cell wall and so are unaffected.*

158. Control of Metabolic Pathways (Page 256)

1. *The end-product of the metabolic pathway may act as the allosteric regulator which attaches to the allosteric site of an enzyme in the pathway and regulates its activity. When production of the regulator declines, it is released from the allosteric site and the pathway is reactivated.*

2. *The end product of the metabolic pathway acts as an inhibitor for an enzyme that controls an early step in the pathway. As the concentration of end-product increases, the pathway becomes less active. In the threonine-isoleucine pathway, the end-product isoleucine inhibits the enzyme in the first step of the pathway.*

159. The Role of ATP in Cells (Page 257)

1. *Cellular respiration.*

2 a. *Adenine*

b. *Ribose*

c. *Phosphate groups*

3. *Adenine and ribose*

4. *Any of:*
Formation of the mitotic spindle and chromosome separation during mitosis, DNA replication, and binary fission, active transport, protein synthesis, movement of cell components or whole cell (protists and bacteria).

160. ATP and Energy (Page 258)

1 a. *The hydrolysis of ATP is coupled to the formation of a reactive intermediate, which can do work. Effectively, the hydrolysis of ATP to ADP + P_i releases energy.*

b. *Like a rechargeable battery, the ADP/ATP system alternates between high energy and low energy states. The addition of a phosphate to ADP recharges the molecule so that it can be used for cellular work.*

2. *Glucose.*

3. *To maintain body temperature.*

161. Cellular Respiration Overview (Page 259)

1 a. *Cytoplasm*

b. *Matrix of mitochondria*

c. *Matrix of mitochondria*

d. *Cristae (inner membrane surface) of mitochondria.*

2. *Glucose + oxygen → water + carbon dioxide + energy (ATP)*

3. *All these pathways have glycolysis as the first step (a universal energy yielding pathway).*

4. *Anaerobic ATP generation in eukaryotes proceeds only as far as the conversion of pyruvate and yields only 2 ATP (from glycolysis). Anaerobic respiration in bacteria proceeds to an ETC and yields much more ATP (although less than aerobic respiration).*

5. *If oxygen enters the fermentation vats, the yeast may respire aerobically and the alcohol (ethanol) would not be produced. Excluding oxygen maximizes the yield of alcohol.*

6. *Glucose → ethanol + carbon dioxide + energy (ATP)*
The difference is that oxygen is not a reactant, and ethanol (or lactic acid) is a product rather than water. The similarities are that glucose is a reactant for both, and ATP is produced (although less for anaerobic).

162. Measuring Respiration (Page 261)

1 a & b.

Time X	Tube 1: Beads alone		Tube 2: Germinating peas				Tube 3: Dry peas and beads			
	Reading at time X (mL)	Difference (correction)	Reading at time X	Difference	Corrected volume (mL)	Respiration rate (mL/min)	Reading at time X	Difference	Corrected volume (mL)	Respiration rate (mL/min)
0	*0.74*	—	*0.66*	—	—	—	*0.81*	—	—	—
5	*0.73*	*0.01*	*0.57*	*0.09*	*0.08*	*0.016*	*0.80*	*0.01*	*0.0*	*0.00*
10	*0.73*	*0.01*	*0.50*	*0.16*	*0.15*	*0.014*	*0.79*	*0.02*	*0.01*	*0.002*
15	*0.73*	*0.01*	*0.42*	*0.24*	*0.23*	*0.016*	*0.79*	*0.02*	*0.01*	*0.00*
20	*0.73*	*0.01*	*0.36*	*0.30*	*0.29*	*0.012*	*0.78*	*0.03*	*0.02*	*0.002*
25	*0.73*	*0.01*	*0.31*	*0.35*	*0.34*	*0.01*	*0.77*	*0.04*	*0.03*	*0.002*

ISBN: 978-1-99-101423-8

c.

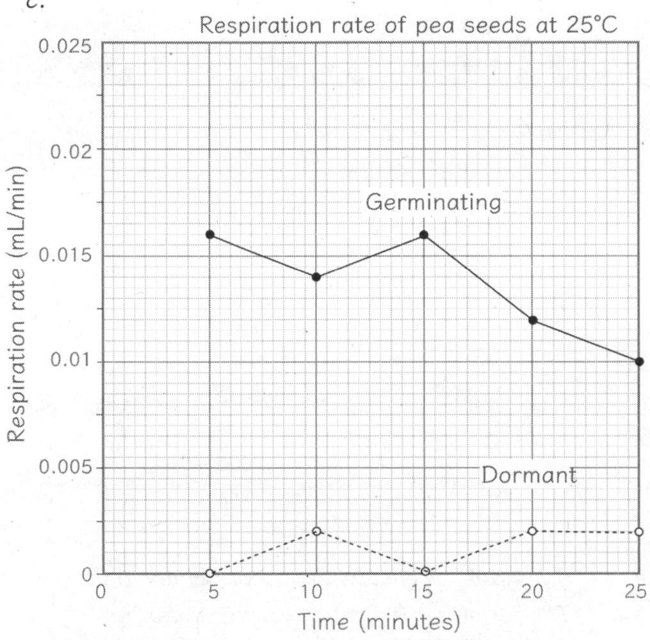

d. The respiration rate is much higher in germinating pea seeds than in dormant seeds.

Respiration rate is constant over time (very slight reduction).

2 a. The KOH absorbs any carbon dioxide produced from cellular respiration.

b. Allows all of the experimental components to equilibrate under the test conditions.

c. This acts as a reference against which to measure the respiration of metabolically active seeds.

d. It is a control. The beads do not respire and occupy the same volume in the tube as the seeds. Any bubble movement in this tube should be deducted from the seed result.

3. Oxygen is being used and the carbon dioxide is being absorbed, so the air pressure in the chamber is dropping, moving the bubble towards the chamber.

4. Changes in temperature and pressure will affect the expansion of air in the tube and so move the bubble. This can be minimized by keeping the tubes in the same constant temperature and pressure.

5. Cellular respiration occurring in ungerminated seeds was very low (barely detectable), but there was some, so the seeds are alive. Cellular respiration occurs at a much higher rate in the germinated seeds as they begin to grow and their metabolic needs are much greater.

6. The plant would have to be kept in the dark because if it was in the light it would photosynthesize and produce oxygen. A true measure of oxygen consumption would not be able to be obtained in the light and the results would not be accurate.

7 a.

Time (minutes)	Distance bubble moved (mm)	Rate (mm/min)
0	0	–
5	25	5
10	65	8
15	95	6
20	130	7
25	160	6

b.

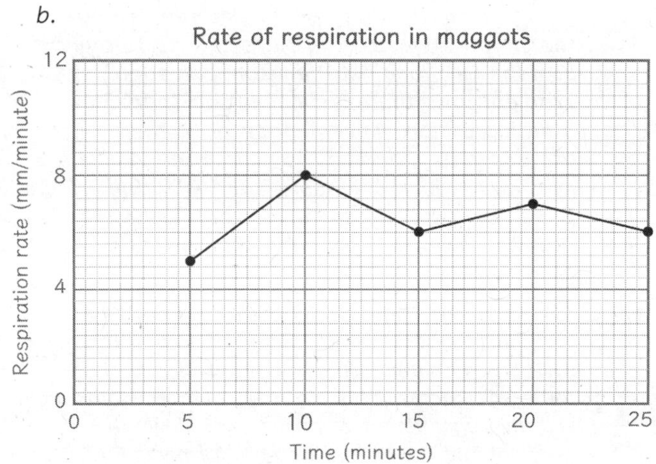

c. After an initial small peak at 10 minutes, the rate of respiration is relatively constant.

163. The Biochemistry of Aerobic Respiration (Page 264)

1 a. Addition of electron from hydrogen - the hydrogen comes from the substrate which loses the hydrogen to become oxidized.

b. The paired reaction with the reduction of NAD^+ or FAD^+ - the substrate loses hydrogen, to become oxidized.

c. The loss of hydrogen from the substrate (also a oxidation reaction if paired with an oxidant - oxidising agent).

d. The addition of phosphate to substrate - this is the paired reaction with ADP to ATP - where the extra P bonds to the ADP.

e. The removal of one or more carbon atoms from a substrate to form a new substrate.

f. Temporary substrates formed during a metabolic pathway that are further reacted to form another product.

164. The Electron Transport Chain and Chemiosmosis (Page 266)

1 a. Substrate level phosphorylation is the direct transfer of a phosphate group from a substrate to ADP (to form ATP).

b. 4 ATP

2 a. Oxidative phosphorylation is the oxidization of glucose through a series of redox reactions resulting in the formation of ATP.

b. 28 ATP

3. $NADH \rightarrow NAD^+ + H^+ + 2e^-$
the H^+ is transported across into inter membrane space using energy from electron pair
$FADH_2 \rightarrow FAD + 2H^+ + 2e^-$
this also occurs in $FADH_2$

4. Hydrogen in the glucose is transferred across to NAD^+ reduction during glycolysis in the cytoplasm and the Krebs cycle in the matrix of the mitochondria and transported across the membrane - into the inner membrane space to build up a proton gradient. The H^+ ions then flow down through the ATP synthase proteins back into the matrix. They combine with oxygen (and electrons) to form water.

ISBN: 978-1-99-101423-8

5. *Energy from the passage of electrons along the chain of electron carriers is used to pump protons (H^+), against their concentration gradient, into the inter membrane space, creating a high concentration of H^+ there. The protons return across the membrane via ATP synthase, which synthesizes the ATP. Elevating the H^+ concentration would result in the H^+ flowing down their concentration gradient via ATP synthase.*
6. *The electrons need to be removed to prevent a build up of charge, and excess H^+ ions in the matrix would impact the proton gradient. Therefore, the oxygen accepts the electrons and ,in combination with hydrogen ions that have flowed down the gradient into the matrix, water is formed.*

165. Anaerobic Pathways (Page 268)

1. *Aerobic respiration requires oxygen and produces a lot of ATP. Fermentation does not require oxygen and uses an alternative H^+ acceptor. There is little ATP produced (the only yield comes from glycolysis).*
2. *In order for glucose to be formed into pyruvate during glycolysis, the NADH needs to be regenerated into NAD^+. This occurs during fermentation. Without this regeneration, glycolysis could not continue.*

166. Investigating Yeast Fermentation (Page 269)

	Cumulative volume of carbon dioxide collected (mL)				
	None	**Glucose**	**Maltose**	**Sucrose**	**Lactose**
0	0	0	0	0	0
5	0	0.6	0.2	0	0
10	0	0.9	0.5	0	0
15	0	1.2	0.8	0.1	0
20	0	2.8	2.0	0.8	0
25	0	4.2	3.0	1.8	0
30	0	5.1	3.6	3.0	0
35	0	7.4	5.4	4.8	0
40	0	10.8	5.6	5.6	0
45	0	13.6	8.0	6.9	0
50	0	16.1	9.6	7.6	0
55	0	22.0	10.4	8.8	0
60	0	23.8	12.1	10.2	0

1. $C_6H_{12}O6 \rightarrow 2C_2H_5OH + 2CO_2$

2 a. *0 cm^3/min*

b. *0.4 cm^3/min*

c. *0.20 cm^3/min*

d. *0.17 cm^3/min*

e. *0 cm^3/min*

3.

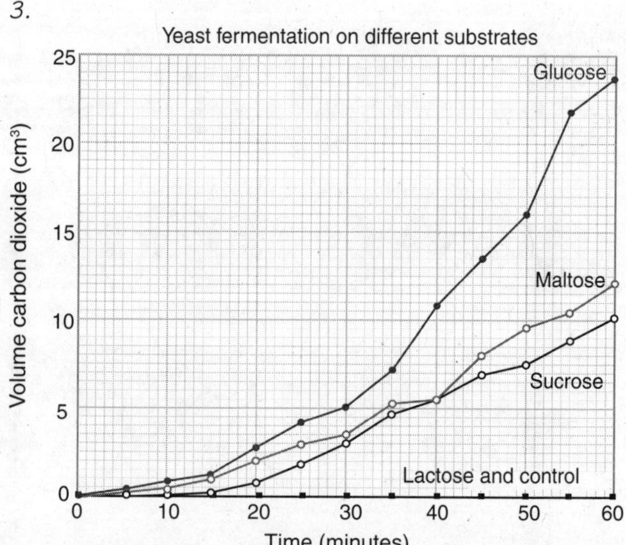

4. *Substrate (this is what you are changing).*

5 a. *The fermentation rates were greatest for glucose, with a CO_2 yield approximately twice that for maltose and sucrose. Maltose and sucrose were similar to each other, while there was no fermentation of lactose.*

b. *CO_2 production was highest when glucose was the substrate because it is directly available as a fuel and requires no initial hydrolysis to use.*

c. *Probably yes. Students may not have expected this. Maltose (glucose-glucose) and sucrose (glucose-fructose) must be hydrolyzed before the glucose is available (the fructose in sucrose must also be isomerized to glucose).*

d. *It is likely that lactose produced no CO2, Yeast lack the enzyme (lactase) to hydrolyze the lactose to galactose and glucose.*

6. *CO_2 production would increase more rapidly.*

167. Photosynthesis Overview (Page 271)

1 a. *Water*

b. *Carbon dioxide*

c. *Photosynthesis*

d. *Oxygen*

e. *Glucose*

168. Separation of Pigments by Chromatography (Page 272)

1. *4 pigments should be seen.*
2. *Students' results will vary but they should see that carotene moves the greatest distance, then chlorophyll a, chlorophyll b, and xanthophyll will move the least.*

169. Pigments and Light Absorption (Page 273)

1. *It is that wavelength of the light spectrum absorbed by a pigment, e.g. chlorophyll, absorbs red and blue light and appears green (reflect this wavelength).*
2. *Accessory pigments (chlorophyll B and carotenoids) absorb light wavelengths that*

ISBN: 978-1-99-101423-8

chlorophyll a cannot absorb and they pass their energy on to chlorophyll a. This broadens the action spectrum over which chlorophyll a can fuel photosynthesis.

3. *Because both oxygen production and carbon dioxide consumption are the result of photosynthesis and not photosynthesis itself - which would be extremely difficult to measure directly.*

4.

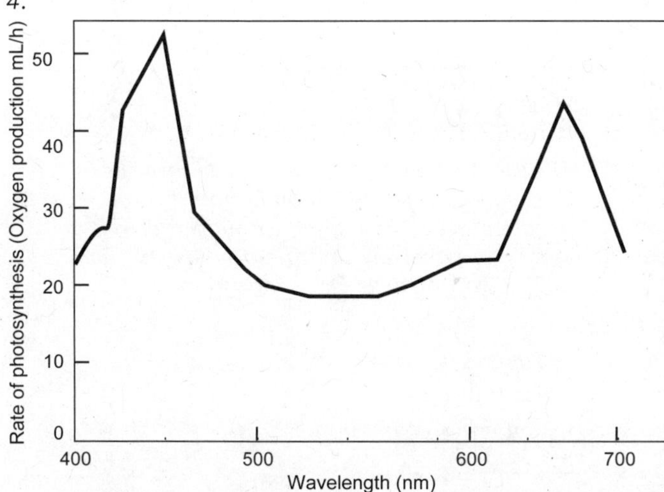

5. *Probably chlorophyll b - there are two sharper peaks - and not the wider production of O_2 between 400-500 nm if there was chlorophyll a as well. Additionally, there is a peak at 650-660 nm which matches chlorophyll b - rather than at 680-690 nm for chlorophyll a.*

6 a. *Students should have drawn graphs with suitable scales and correctly labelled axes, using the data in the table.*

b. *Most likely blue light - as closet to that value in O_2 and CO_2 values - but with some additional colours as well.*

c. *Some residual photosynthesis could occur (light independent) if substrates were already present - alternatively - it could be an error in the measuring equipment.*

170. Factors Affecting Photosynthesis (Page 275)

1 a. *Increasing the amount of CO_2 available to a plant will increase the rate of photosynthesis - up to 80 mm^3 CO_2/ cm^2/hr*

b. *Increasing the amount of light available to a plant will increase the rate of photosynthesis - up to a certain intensity*

c. *Increasing the temperature available to a plant will increase the rate of photosynthesis - but not beyond a certain temperature when damage can occur to a plant (i.e. above 40 degrees)*

2. *The photosynthetic rate is determined by the rate at which CO_2 enters the leaf. When this declines because of low atmospheric levels, so does photosynthetic rate. However, once there is sufficient CO_2 then light intensity becomes the determining limiting factor. Temperature has some effect - but not as significant as the other two factors.*

Investigation 9.5

Distance (cm)	Light intensity (lx)	Bubbles counted in three minutes	Bubbles per minute	Volume (mL)
50 cm	*5*	*0*	*0*	
45 cm	*13*	*6*	*2*	
40 cm	*30*	*9*	*3*	
35 cm	*60*	*12*	*4*	
30 cm	*95*	*18*	*6*	
25 cm	*150*	*33*	*11*	
20 cm	*190*	*35*	*11.67*	

3.

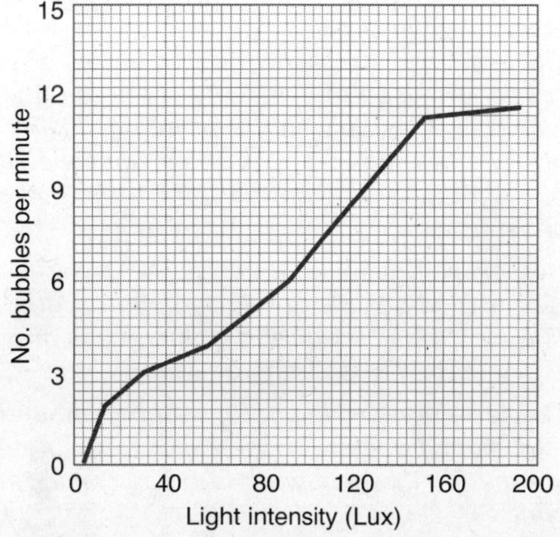

4. *Measuring lux produces specific quantitative data independent of the distance. Light can be affected by shadows etc so inferring intensity from distance is inaccurate.*

5. *The gas was oxygen.*

6. *Counting bubbles is inaccurate because the bubble may be of different volumes so will not infer total volume.*

7. *Depends on student's original hypothesis - most likely that increase in light intensity would increase rate of photosynthesis -so agree.*

171. Details of Photosynthesis (Page 278)

1 a. *Light dependent phase.*
Location: in the thylakoid membranes of the grana.

b. *Light independent phase.*
Location: in the stroma.

2 a. *Rubisco catalyses the first step of the Calvin cycle.*

b. *Rubisco makes up half the protein content of chloroplasts. As plants are very abundant on Earth, and each plant leaf cell contains 50-100 chloroplasts, the total amount of Rubisco is very high.*

3 a. *CO_2 origin: from the air. Fate: incorporated into glucose.*

b. *O_2 origin: from splitting of water. Fate: released as O_2 gas.*

ISBN: 978-1-99-101423-8

c. *H_2 origin: from splitting of water. Fate: is incorporated into glucose and water.*

172. Light Dependent Reactions (Page 279)

1. *Carries protons (H^+) from the light dependent phase to the light independent reactions.*
2. *Chlorophyll molecules trap light energy and produce high energy electrons. These are used to make ATP and NADPH. The chlorophyll molecules also split water, releasing H^+ for use in the light independent reactions and liberating free O_2.*

3 a. *The grana (thylakoid membranes) of the chloroplast*

b. *The light dependent reactions require light to proceed. The light dependent phase generates ATP and reducing power in the form of NADPH. The electrons and hydrogen ions come from the splitting of water.*

4. *When the light strikes the chlorophyll molecules, high energy electrons are released by the chlorophyll molecules. The energy lost when the electrons are passed through a series of electron carriers is used to pump H^+ across the thylakoid membrane, establishing a proton gradient. The protons flow back across the membrane via ATP synthase, generating ATP from ADP and phosphate.*

5 a. *Non-cyclic phosphorylation is the generation of ATP using light energy during photosynthesis. The electrons lost during this process are replaced by the splitting of water.*

b. *The term non-cyclic photophosphorylation is also (commonly) used because it indicates that the energy for the phosphorylation is coming from light.*

173. Light Independent Reactions (Page 281)

1.

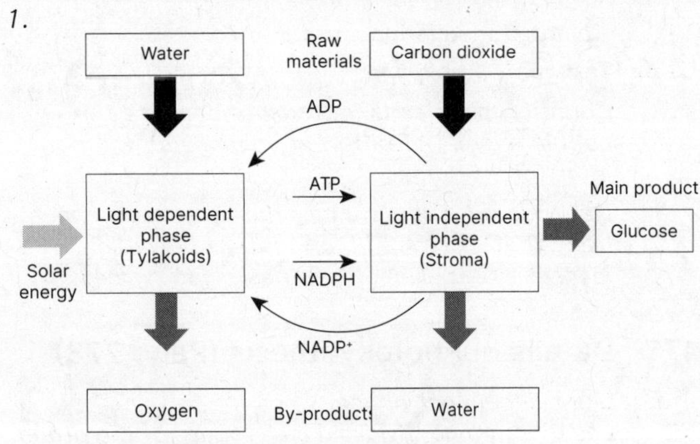

174. Did You Get It? (Page 282)

1. *The active site is where the specific substrate attaches - induced fit is when the substrate modifies the shape of the enzyme.*
2. *Digestion takes place inside the lumen of digestive organs - before digestion, the macromolecules are too large to fit inside cells - extracellular enzymes need to facilitate catabolic reactions to break them into smaller units.*
3. *Linear pathways start with a substrate/s and through a series of reactions, produce a final product. Cyclic pathways have the final product as the starting substrate of the next cycle - although they also consist of a series of enzyme facilitated reactions - and tend to be more complex than linear pathways.*

4 a. *2*

b. *28*

c. *Glycolysis*

d. *Krebs cycle*

e. *Electron transport chain*

f. *CO_2*

g. *CO_2*

h. *Water*

5. *The H+ ions (protons) are moved into the intracellular space and build up a proton gradient. They flow back through the ATP synthase passively and their movement enables ATP to form from ADP - phosphorylation.*

6 a. *Water*

b. *Oxygen*

c. *Carbon dioxide*

d. *(Numbering error; no letter d)*

e. *Glucose*

f. *Photolysis*

g. *Thylakoid*

h. *Calvin cycle*

i. *Stroma*

7. *RuBisCo catalyzes the reaction that splits CO_2 and joins it with ribulose 1,5-bisphosphate. It fixes carbon from the atmosphere.*

Theme C: Interaction and Interdependence

Chapter 10: Cells

175. Signals and Signal Transduction (Page 285)

1 a. *Reception: The signal molecule binds to the receptor on the cell surface.*

b. *Transduction: The activated receptor triggers a series of biochemical events within the cell (a signal cascade).*

c. *Response: The signal cascade results in a specific response in the target cell.*

2. *Only cells with the specific receptor molecules on their surface will respond to specific signal molecules.*

3 a. *The autoinducer acts as a signal for luminescence when a critical level is encountered (indicating a certain population level) and the luminescence pathway is activated.*

b. *The level of autoinducer is related to the population density. The higher the level of autoinducer, the denser the population of bacteria.*

c. *Population density information is important in regulating bacterial behaviour. Luminescence may only occur in the safety of a large bacterial colony. Virulence factors may only switch on when the population is dense enough to ensure*

ISBN: 978-1-99-101423-8

infection of a host.

4. *Bobtail squid concentrate bacteria in a special organ. Concentrating the bacteria induces the expression of luminous proteins. The squid uses them to produce light to help in camouflage.*

5 a. *2*

b. *3*

c. *1*

d. *4*

176. Signal Molecules (Page 287)

1. *Neurotransmitters are short ranged signalling chemicals that are passed between neurons. Their effects are short lived. Hormones are long range signalling molecules that are passed between potentially distant cells. Their effects are long ranged.*

2 a. *Hormone*

b. *Neurotransmitter*

3. *Cytokines are important in coordinating the response of cells in the immune system.*

4. *Cytokines are produced by a range of different cells and circulate in more variable concentrations than hormones. Hormones are produced by endocrine cells and circulate in low concentrations.*

5. Ca^{2+}

6. Ca^{2+} *can move through gap junctions.*

7. *Pheromones are secreted into the external environment to influence other organisms, not control the physiology of the animal producing them.*

8. *Nitric oxide is a very small molecule that can easily cross the plasma membrane. Therefore, storing it in a cell would be very difficult (it would diffuse out of any storage vesicle). Making it, on demand, from precursors that can be stored is a better option.*

9. *Having both excitatory and inhibitory neurotransmitters gives a greater range of control for neurons, as these neurotransmitters are able to reverse the actions of each other.*

10. *Steroid hormones are able to pass through the plasma membrane and bind to receptors inside the cell. Peptide hormones bind to receptors on the outside of the cells.*

11. *Having many different signalling molecules gives the ability to finely regulate multiple metabolic and cellular pathways. It also makes sure that signals are specific and not ambiguous.*

177. Communication Over Short and Long Distances (Page 289)

1. *In animals, local regulation is used to transmit signals between neurons. Neurotransmitters from the signalling cell are released into the synapse and received by the receptors on the target cell, which then triggers a response (nerve impulse).*

2. *Cells communicate over long distances using hormones. These are produced by endocrine cells and are transported in the blood to target cells which may be some way from where the hormones was produced.*

178. Transmembrane and Intracellular Receptors (Page 290)

1. *Extracellular receptors bind signal molecules outside the cell (on the cell's surface). The signal molecule does not cross the plasma membrane. Intracellular receptors bind signal molecules that have passed through the plasma membrane.*

2. *Ligands that bind to intracellular receptors are hydrophobic and are able to cross the plasma membrane. Ligands that bind to transmembrane receptors are hydrophilic and cannot cross the plasma membrane.*

179. Transmembrane Receptors for Neurotransmitters (Page 291)

1. *Acetylcholine binds to the acetylcholine receptor, a ligand gated ion channel, causing the channel to open. Sodium ions flow through the channel. This causes a local depolarization of the membrane, called an end plate potential. If the end plate potential is large enough, it causes voltage gated ion channels to open, letting more* Na^+ *into the cell, resulting in depolarization and an action potential.*

180. G-Coupled Protein Receptors (Page 292)

1. *A first messenger binds to and activates a G-protein linked receptor, which activates adenylate cyclase. This catalyses the synthesis of cAMP. cAMP initiates a phosphorylation cascade that leads to a cellular response. The role of G proteins, therefore, are to act as molecular switches, regulating the activity of a signal transduction pathway.*

2. *A water soluble first messenger cannot cross the plasma membrane as it is polar and repelled by the non-polar nature of the phospholipid tails in the plasma membrane. It must interact with a receptor on the membrane and activate a second messenger inside the cell.*

181. Tyrosine Kinase Activity in Receptors (Page 293)

1 a. *Endocrine*

b. *Insulin is produced in the pancreas and travels in the bloodstream to influence the cells of the body.*

2. *Once a signal is received, tyrosine kinase autophosphorylates (adds phosphate to itself). It then phosphorylates other molecules, activating them and beginning signal cascades within the cell, which leads to glucose entering the cell.*

3. *In a signal cascade, the sequence (chain reaction) of phosphorylations results in the activation of an increasing number of other proteins. Thus, for two signal molecules (insulin), many proteins are activated and many Glut4 secretory vesicles (each producing transporters) are made.*

ISBN: 978-1-99-101423-8

182. Intracellular Receptors (Page 294)

1. *Steroid hormones are able to cross the plasma membrane and directly attach to a steroid receptor (transcription factor) which can enter the nucleus and induce transcription of specific genes.*
2. *Steroid hormones are able to interact directly with the DNA and are thus able to control when genes are expressed during the developmental process.*
3. *Oestradiol (female) and testosterone (male).*

183. Feedback and Hormones (Page 295)

1. *FSH stimulates the growth of ovarian follicles.*

   LH stimulates ovulation and development of the corpus luteum.
2. *Oestrogen is secreted in response to FSH. At a certain level, the oestrogen inhibits further FSH release so that the development of more follicles is prevented. Oestrogen levels fall and the next stage of the cycle proceeds.*

184. Neuron Structure and Function (Page 296)

1. *A neuron processes and transmits information as electrical impulses from the point of stimulus to the point of reception.*
2. *A larger axon diameter and increased myelination.*
3. *Myelin prevents ions from entering or leaving the axon along myelinated segments and prevents leakage of charge across the neuron membrane. It forces the action potential to jump from one node to the next.*
4. *Faster conduction speeds enable more rapid responses to stimuli.*

185. The Nerve Impulse (Page 297)

1 a. *The resting potential is the voltage difference across the membrane (between the inside and outside of the cell). In a non transmitting neuron, it is -70mV.*

b. *An action potential is a self-regenerating depolarization that allows excitable cells (such as muscle and nerve cells) to carry a signal over a (varying) distance.*

2. *The resting potential is maintained by the sodium/potassium pump. Three sodium ions are pumped out of the cell for every two potassium ions pumped into the cell. This creates the potential difference across the membrane.*
3. *An action potential is propagated by changes in local currents/voltage. Depolarization in one region causes a voltage change in the neighbouring region which opens voltage gated ion channels and so causes a depolarization.*

186. The Speed of Nerve Impulses (Page 298)

1.

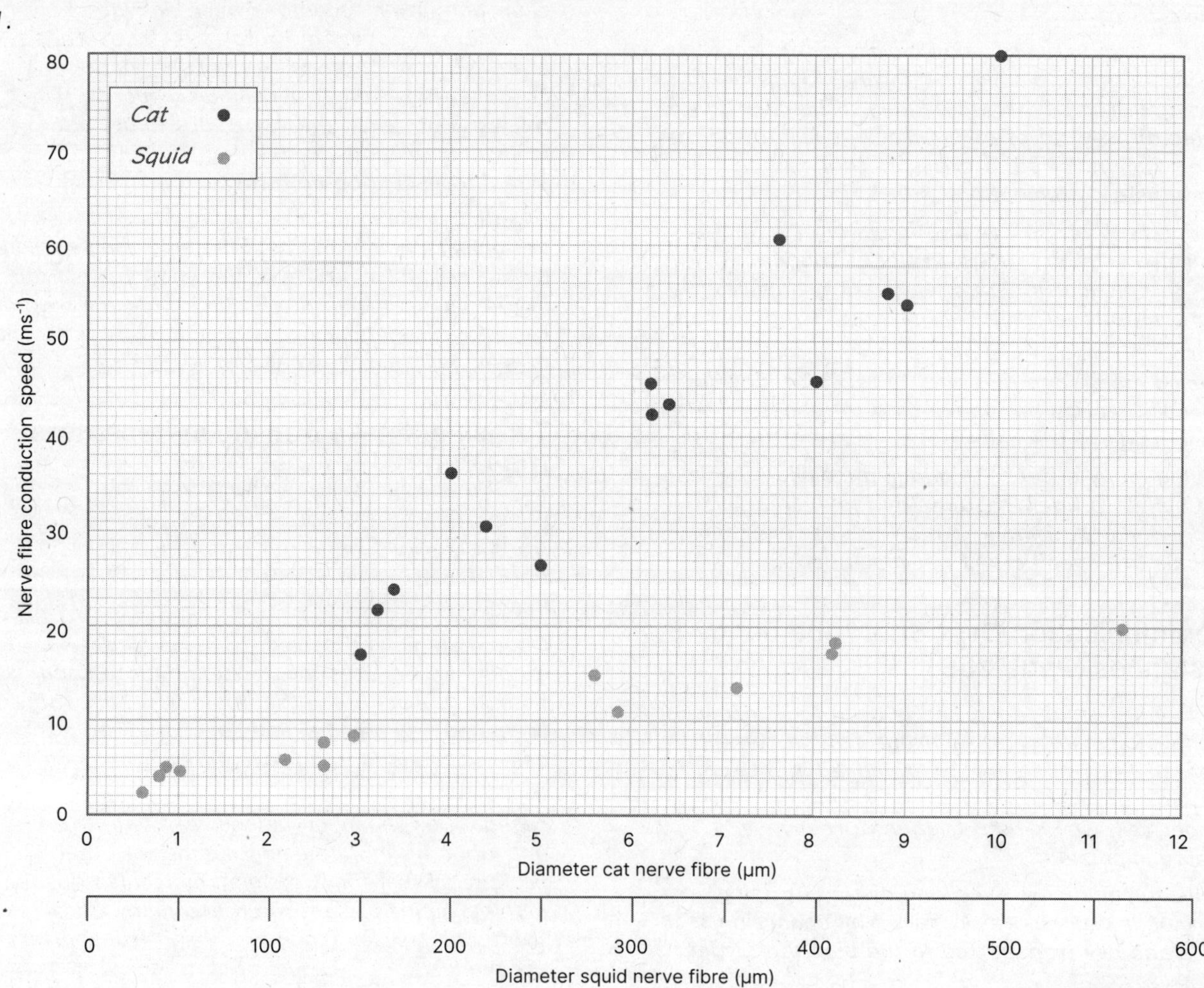

ISBN: 978-1-99-101423-8

2 a. 0.87

b. 0.93

3. Myelination is more important for speed of transmission.

187. Chemical Synapses (Page 299)

1 a. A synapse is a junction between the end of one axon and the dendrite or cell body of a receiving neuron. A synapse can also occur between the end of one axon and a muscle cell.

b. The type of neurotransmitter they release, i.e. acetylcholine.

2. Arrival of a nerve impulse at the end of the axon causes a calcium influx. This induces the vesicles to release their neurotransmitter into the cleft.

3. Delay is caused by the time it takes for the neurotransmitter to diffuse across the synaptic cleft.

4. The amount of neurotransmitter released influences the response of the receiving cell (response strength is proportional to amount of neurotransmitter released).

188. Electrical Impulses in the Nerve (Page 300)

1. Nothing (no propagation of signal)

2. In myelinated neurons, voltage gated ion channels are found only at the nodes of Ranvier. In nonmyelinated neurons, the channels are found all along the neuron axon.

3. One response takes 4 ms so in one second the nerve cell can respond 250 times.

189. Chemicals at Synapses (Page 301)

1. Block transmission of the neurotransmitter by binding to the receptor permanently, block reuptake of the neurotransmitter by binding to the reuptake channel, mimic a neurotransmitter by binding to the receptor (reversible).

2 a. Neonicotinoids mimic the action of acetylcholine on cholinergic receptors but they are not broken down by the enzyme acetylcholinesterase meaning they can bind to the cholinergic receptors permanently and block the transmission of nerve signals.

b. Neonicotinoid pesticides have been restricted in many countries because they are a broad spectrum insecticide and have been shown to harm honey bees which are vital for crop pollination.

3. The cocaine molecule blocks the dopamine transporter. Dopamine builds up in the synaptic cleft. The dopamine therefore continues to act on the dopamine receptor, prolonging its effect.

190. Summation at Synapses (Page 302)

1 a. The additive effect of presynaptic inputs (impulses) in the postsynaptic cell.

b. Spatial summation refers to summation of impulses from separate axon terminals arriving simultaneously at the postsynaptic cell. Temporal summation refers to the arrival of several impulses from a single axon terminal in rapid succession.

2. Hyperpolarlization takes the membrane potential further away from the threshold potential (makes the membrane potential more negative), thus the nerve is less likely to produce a signal until an excitatory signal is received.

191. The Perception of Pain (Page 303)

1. Nociceptors respond to stimuli that cause damage to the body by sending signals to the brain that are interpreted as pain.

2. Damaging mechanical, chemical, thermal stimuli.

3 a. Thermal/ heat sensitive ion channels.

b. Mechanically gated ion channel.

c. Mechanically gated ion channel.

4. Nociceptors are polymodal because they are able to respond to multiple types of stimuli.

5. You feel heat where capsaicin has been encountered (e.g. around the mouth when eating spicy food).

192. What is Consciousness? (Page 304)

1. Emergence is the property of small building blocks working together to produce a large object with more complex properties than just the sum of the smaller units. E.g. consciousness is produced by billions of non conscious cells working together.

2 a. Studies show consciousness is not limited to one part of the brain and that communication between neurons is important. Thus, consciousness could be everywhere and nowhere in the brain, making finding its exact origin very difficult.

b. Consciousness is, to a degree, subjective- how do we define consciousness and at what point is something conscious? Different organisms may have different experiences of consciousness, making our point of view somewhat misleading.

193. Did You Get It? (Page 305)

1 a. A

b. They are the wrong shape; they will not fit the receptor.

c. If all cells reacted to all signals, there would be an inefficient use of materials in the cell. Cells that would not need to react would do so.

2. The self propagating depolarization of the plasma membrane in response to the threshold potential being reached.

3 a. Axon branches

b. Dendrites

c. Cell body (soma)

d. Axon

4 a. Nerve impulses move down a myelinated axon by depolarization only at the nodes of Ranvier.

b. It increases the speed of conduction.

5. B - Membrane depolarization (due to rapid Na+ entry across the axon membrane.

ISBN: 978-1-99-101423-8

D - Hyperpolarization (an overshoot caused by the delay in closing of the K+ channels).

E - Return to resting potential after the stimulus has passed.

C - Repolarization as the Na+ channels close and slower K+ channels begin to open.

A - The membrane's resting potential.

Theme C: Interaction and Interdependence

Chapter 11: Organisms

194. Biological Organization (Page 308)

1. *A life function such as reproduction, that requires integration of components at a lower level, but is not possessed by the individual components themselves.*

2 a. *The organelle level.*

b. *The organism.*

c. *The chemical level (DNA).*

195. Integration in Nervous and Hormonal Signalling (Page 309)

	Nervous control	Hormonal control
Communication	Impulses directly between cells across cell to cell junctions.	Hormones in blood.
Speed	Very rapid.	Relatively slow.
Duration	Short term.	Longer lasting.
Target pathway	Through nerves to specific cells.	Carried in blood throughout body to target cells.
Action	Causes glands to secrete or muscles to contract.	Causes changes in metabolic activity.

196. Integration in Materials and Energy Transport (Page 310)

1.

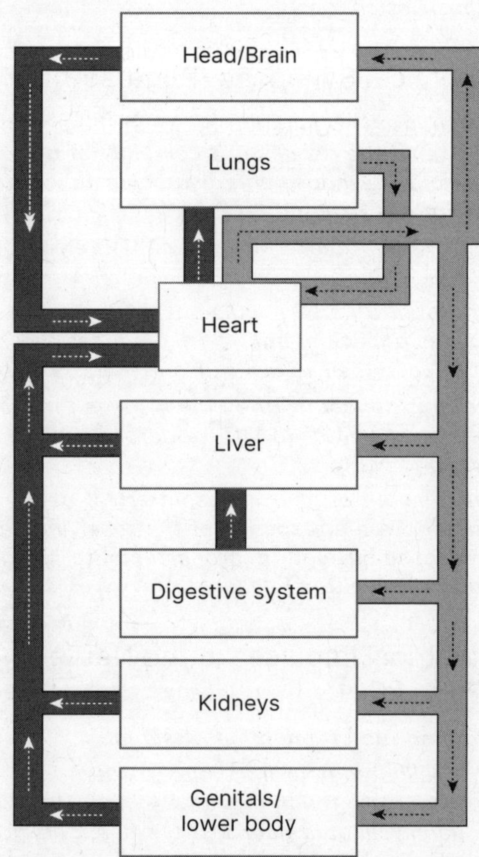

197. Information Integration in the Brain (Page 311)

1. *Information is received from all areas of the body via sensory neurons. This information is then processed and appropriate 'instructions' are sent out for activating a response to specific tissues or cells, typically via motor neurons.*

2. *Receives information from sensory receptors in the skin, viscera, and muscles - awareness of the location and intensity of pain, temperature, and touch.*

3. *The primary motor area controls voluntary muscle movement, i.e. to generate nerve impulses that control muscles.*

4 a. *Occipital lobe.*

b. *Primary gustatory area.*

c. *Motor and sensory speech areas (Broca's and Wernicke's areas respectively).*

d. *Frontal lobe of the cerebrum.*

5. *Short-term memories are temporary recollections from sensory input and are stored in neurons in the prefrontal cortex. These memories are lost quickly if not reinforced. Long term memories are the result of repeated stimuli and are held in the hippocampus, amygdala, neocortex, or cerebellum, depending on memory type.*

6 a. *Conscious memory - formed by real-world experiences or traditional 'learning' of information - recalled with intentional thought.*

b. *Episodic memory occurs when we experience*

ISBN: 978-1-99-101423-8

real-world events, perceived by one or more senses, and then stored it for recall in our memory centres. Semantic memory is created when 'learning' information or facts, but not experiencing the events. Sometimes other inputs, such as sound or smell, can be stored alongside if occurring simultaneously.

7. *Repetition of actions or input. Each time an event or action is repeated, it strengthens neural connections, making the memory stronger - also requires sleep to 'embed' the memories. Storing associated emotions at the time of the memory also strengthen them.*
8. *The learning of skills, such as driving or walking, can result in memories being stored in the cerebellum. They can then be recalled and actioned unconsciously - so we can perform them without thinking about how we do them.*

198. The Autonomic Nervous System (Page 313)

1 a. *The processes occur without the input of conscious thought; they occur in the 'background' in responses to internal stimuli (often involved in homoeostasis).*

b. *The process is a life function that needs to occur even when sleeping (unconscious); also, there are many integrated processes that occur at a lower level and they all need to be carefully synchronized.*

2. *The parasympathetic system increases salivation, insulin, and enzyme release when food moves through; it also increases peristalsis of the stomach and intestines to move food through, and the liver increases sugar storage. When there is no food moving through, the sympathetic system inhibits or decreases these processes.*

3 a. *Neural connections from most regions of the spinal cord. This system is a slower acting system to mostly suppress or inhibit unconscious processes in the autonomic system.*

b. *Neural connections primarily from the cranial and sacral regions of the spinal cord. A fast acting system that stimulates tissues and organs to secrete, contract. The exception is the heart rate - where this system promotes slowing of the heart rate (for high blood pressure etc), while the sympathetic system promotes increasing the heart rate.*

4. *Most sympathetic postganglionic nerves release norepinephrine which enters the bloodstream and is deactivated slowly. In contrast, parasympathetic nerves release acetylcholine which is rapidly deactivated at the synapse with short-lived localized effects.*

5. *A sudden decrease in blood pressure would result in the sympathetic NS increasing the constriction of the coronary blood vessels and subsequent increase in the rate and force of contraction of the heart. As a result, blood pressure would rise. Increase in arterial pressure causes reflex stimulation of the cardioinhibitory centre through the parasympathetic division, slowing heart rate and decreasing arterial BP to normal.*

6. *Bladder emptying is under reflex control and is stimulated by stretching of the bladder wall. Stretching causes both a conscious desire to urinate and an unconscious reflex contraction of the bladder wall and relaxation of the internal urethral sphincter. The conscious part of the brain also sends impulses to relax the external urethral sphincter. Because both conscious and unconscious controls are involved, urination can be voluntarily stopped and started at will.*

199. Neural Inputs and Outputs (Page 315)

1. *Sensory neurons typically have their receptors located on the periphery of the body, and the neuron/s then travel to the brain, usually via the spinal cord (they are receiving stimuli). The motor neurons begin from the brain and travel to the tissue / muscle to allow signals for contraction (they are sending stimuli).*
2. *The left half of the body, in particular the muscles, are controlled by the right hand side of the cerebral cortex, and vice versa. Nerve pathways cross over at the medulla region of the brain.*
3. *Repetitive movements learnt as skills can be controlled by the cerebellum 'automatically' without involving the longer process of routing through the cortex.*
4. *Cortex. Sometimes a change in the environment requires a conscious decision to alter a usually automatic motion. For example, walking on a flat path then suddenly encountering a hole or rock - consideration of how to avoid.*
5. *Nerve transmitted information gathered from receptors to indicate the position of the body in relation to the external environment: proprioceptive receptors in muscles, joints, and skin and vestibular receptors in the inner ear.*

6 a. *In the somatosensory system, sensory information passes through the cerebellum and crosses over in the medulla before travelling to the cerebral cortex on the opposite side, where the information is interpreted (integration).*

b. *The cerebellum receives information from higher brain centres about what the muscles should be doing and compares this to information it receives about what the muscles are actually doing. It then provides corrective feedback to smooth motor responses.*

7. *The cerebellum, because it is constantly providing corrective feedback to coordinate muscle movements in response to information from higher brain centres.*

200. Structure of Nerves (Page 317)

1 a. *Myelination increases the speed of impulse conduction. The impulses jump from 'gap' to 'gap' instead of moving along the entire axon.*

b. *Oligodendrocytes*

c. *Schwann cells*

d. *Neurons in the peripheral nervous system frequently have to transmit over long distances so speed of impulse conduction is critical to efficient function.*

ISBN: 978-1-99-101423-8

2. *Muscles, including cardiac, smooth, and skeletal; and secretary glands, such as those in the stomach.*

201. Reflexes (Page 318)

1. *Conscious thought would slow down the response time. The adaptive value of reflexes is in allowing a very rapid, automatic response to a stimulus.*

2 a. *A monosynaptic reflex involves only one CNS synapse (two neurons) whereas a polysynaptic reflex involves two or more (involves a relay or interneuron).*

b. *The monosynaptic reflex because fewer synapses are involved, therefore less delay in impulse transmission.*

3. *A startle reflex may have reduced the effect of nuisance pests or predation (throwing out arms may scare off a potential predator or shift an annoying fly). A grasp reflex would help the infant cling onto its mother when the mother is moving.*

202. Sleeping and Waking (Page 319)

1 a. *Hypothalamus - superchiasmatic cells.*

b. *The biological clock is a structural feature in the hypothalamus (keeps a cycle around 24 hrs, 11 minutes) - the circadian rhythm is the behavioural and physical changes induced by the biological clock.*

2. *Melatonin is a hormone secreted by the pineal gland, and controlled by the SCN, as part of the biological clock. The melatonin decreases body temperature and promotes sleepiness as daylight fades. The SCN suppresses melatonin once daylight appears and the body becomes more wakeful, and body temperature rises once more.*

203. Control of the Endocrine System (Page 320)

1. *Information from receptors arrives at the hypothalamus from both nervous and hormonal signals. Further hormones are then released from the hypothalamus that either result in release or inhibition of hormones from the pituitary gland. In turn, those hormones released from the pituitary gland determine the action of other endocrine glands in the body.*

2. *Stimulus from the hypothalamus causes the release of epinephrine (a type of catecholamine) from the adrenal glands, which prepares the body for action by increasing heart rate and blood pressure, and mobilizing glucose. Blood flow to the brain and muscles is increased and metabolic rate increases. The epinephrine binds to cAMP receptors on muscle cells, allowing for increased contractions.*

204. Control of Heart Rate (Page 321)

1 a. *Heart rate increases.*

b. *Heart rate increases.*

c. *Heart rate increases.*

2. *These effects are mediated through the cardiovascular centre (sympathetic output via the cardiac nerve).*

3 a. *Cardiac nerve*

b. *Vagus nerve.*

4. *Increased stretch in the VC indicates increased venous return so cardiac output must increase to cope with the greater volume. Increased stretch in the aorta indicates increased cardiac output and heart rate decreases.*

205. Control of Ventilation rate (Page 322)

1. *It is controlled by the respiratory centre in the medulla, which sends rhythmic impulses to the intercostal muscles and diaphragm to bring about normal ventilation.*

2 a. *Innervates the diaphragm (which contracts and moves down in inspiration).*

b. *Innervate the intercostal muscles (and nerves) to bring about ribcage movements.*

c. *Sensory portion carries impulses from stretch receptors in the bronchioles to the respiratory centre to inhibit inspiration.*

d. *The inhibition of the inspiratory centre to end the breath inward.*

3 a. *Low blood pH increases rate and depth of breathing.*

b. *Sensory information from aortic and carotid chemoreceptors is sent to the respiratory centre, which mediates the increase in breathing rate.*

c. *Blood pH is a good indicator of high CO_2 levels (therefore to increase respiratory rate to remove the CO_2 and obtain more oxygen).*

206. Control of Peristalsis (Page 323)

1. *Ingestion involves food and liquid entering the mouth and swallowing. Egestion involves removal of waste from the anus. Both processes are under CNS so they can be consciously controlled (some foods or liquids may not be swallowed by choice). Egestion occurs when the environment is chosen. However, peristalsis occurs 'automatically' and unconsciously once food boluses have entered the digestive system - and can occur during activity or sleep. Therefore, this process is under the autonomic ENS.*

2. *The peristalsis movement involves the coordination of several muscle groups in sequence. This needs to be performed continuously during both activity and sleep so food boluses do not stagnate in the tract.*

3. *Ingestion: to prevent unwanted food being swallowed - either distasteful or too large - and avoiding choking. Thought processes in the CNS can consider when and what to swallow.*

Egestion: the CNS is required so consideration can be given to the correct location and time. Important in animals especially, when a safe area is needed for the process.

207. Investigating Phototropism (Page 324)

1. *To observe bending that occurs without light so that value can be subtracted from the measurements of plants exposed to light.*

2. *Quantitative: measurements to obtain values - for example the angle of bend in seedling stems. Qualitative: observations recorded, for example,*

ISBN: 978-1-99-101423-8

the shape of the seedlings after exposure to light (plus control).

3. *Report to include:*

Hypothesis: If plants grow towards a light source the seedlings will either grow straight or bend depending on where the light source is.

Method: Can be summarized from the worktext method provided.

Dependent variable: Angle of stem bend.

Independent variable: Position of the light hole in relation to the seedling.

Precision and accuracy: Take measurements from the same position on the stem and use the same instrument for measuring. Removing seedlings that do not properly grow - not including in data. Have the same student take the measurements each time.

Reliability: Use at least three seedlings per trial. Calculate average measurements and exclude any outliers, e.g. stunted or sick plants. Ensure the slots cut are the same size for each box. Students may reference the purpose of the control to show the results are in response to the light source.

Qualitative diagrams: Simple line drawings similar to those shown in the worktext.

Quantitative data: Record data in tables, allowing for means to be calculated.

Conclusion: Students should have seen that the seedlings grew towards the light source and the angle of the stem varied depending on the position of the light source. Seedlings in group A showed the largest angle as the seedlings bent horizontally towards the light source. Seedlings in group B grew upwards, but the stems had a slight angle compared to groups A and C as the light source was vertical but off-centre. Seedlings in group C grew vertically towards the overhead light source,showing little to no bend in their stems. The control group D may have showed no growth or died due to the lack of a light source. Students can conclude that light (and not any other factor) caused the differences in stem angles observed.

208. Phytohormones as Signalling Chemicals (Page 325)

1. *Gibberellins are strong promoters of elongation in stems by stimulating both cell division and cell elongation. In seeds, gibberellins are responsible for breaking dormancy and stimulating the growth of the embryo and emergence of the seedling.*
2. *In response to water stress, ABA stimulates the closing of stomata. Note: It does this by stimulating the loss of K^+ from the guard cells (via Ca_2^+ as second messenger).*
3. *ABA is involved in processes associated with ageing and cessation of growth, e.g. fruit fall and seed dormancy, It generally opposes the effects of growth promoting hormones, such as the cytokinins.*

209. Auxin and Phototrophism (Page 326)

1. *A plant response controlled by hormones (auxin) that allows a plant to bend towards a light source by elongating cells in the stem on the shaded side.*
2. *Auxins move in one way through auxin efflux carriers = protein channels embedded in the cell membrane unevenly distributed - and so move from one cell to another to set up a gradient in plant tissue. The pH environment concentrated in one end of the cell prevents the auxin from moving through the channels in the opposite direction.*
3. *As the auxin concentration increases in the cell, this increases the number of hydrogen ions pumped into the apoplast (the space between cell walls of adjoining cells) through ion pumps. The increased pH then activates expansins to loosen cellulose and allow cell walls to stretch.*

4 a. *Point A: No growth*

Point B: Elongation due to increased auxin levels

b. *Side B - higher auxin levels as they have moved through the auxin efflux carriers in one direction way from light to increase in concentration on the shaded side.*

5. *The auxin hormone needs the stimulus of light in order to initiate movement through the auxin efflux carrier to bring about a concentration gradient. Without light (clocked from tin-foil), then the auxin does not move from cell to cell.*

210. Interaction of Plant Hormones (Page 328)

1. *A plant requires a strong and deep vertical main root (to reach deeper water supplies and anchor the plant) - and auxin encourages vertical root development. The plant also needs a broad surface area of roots (and root hairs) that expand out from the plant footprint - so cytokinin encourages horizontal root growth.*
2. *The plant can grow upwards due to auxin acting on the stem meristem - but also requires differentiated lateral development induced by cytokinin - especially if the plant loses its meristem.*
3. *If the meristem /apex of the plant stem is removed or lost, so too is the concentrated auxin hormone at its tip. This allows the concentration of the cytokinin to become more significant and drive the development and differentiation of the lateral buds, making the plant grow more bushy. The auxin is the antagonist and, in high concentrations, as is found in the apex, this hormone reduces the effect of cytokinin.*

211. Positive Feedback Loops and Fruit Ripening (Page 329)

1. *Fruit can be picked when still unripe - allowing transport of the fruit and storage over a longer period of time. Ethylene is then used to ripen the fruit before it is sold, so that the extent of ripeness is timed to when the fruit is bought and consumed. It can also be used to uniformly ripen fruit that grows in a bunch - such as bananas and tomatoes - so that all fruit is ripe and ready for sale at the same time.*

ISBN: 978-1-99-101423-8

212. Pathogens and Disease (Page 330)

1 a. *Pathogen*
b. *Cellular pathogen*
c. *Non-cellular pathogen*
d. *Prokaryote*
e. *Eukaryote*
f. *Bacteria*
g. *Protist*
h. *Parasitic worm*
i. *Fungi*
j. *Virus*
k. *Prion*

2. *Cellular pathogens are living organisms which can cause disease. They have the cellular machinery needed to carry out their life processes independently. Non-cellular pathogens are classed as non-living pathogens. They cannot metabolize independently because they do not have their own cellular machinery (they use the host cell machinery).*

3. *The source of the outbreak is pump 3. This is the centre of the cases in the outbreak. Cholera is shown linked to water pollution. Districts with low elevation, taking polluted water from further down the Thames have much higher rates of cholera. Prevention would be to stop people taking water from pump 3 (removed from the pump).*

4. *No. The outbreak was already declining in number before the pump handle was removed. (John Snow's work was more important in establishing a method for finding and linking the source of pathogens).*

5. *Student answers will vary depending on the disease they have chosen. Some information for childbed fever is provided.*

Disease: Childbed fever, also called puerperal fever, is a disease affecting some women within three days of giving birth.

Cause: Several factors may cause childbed fever, but infection of the uterus caused by the bacteria <u>Streptococcus pyogenes</u> or <u>Staphylococci</u> spp are a common cause. If left untreated, sepsis can occur and death may result.

In the 1800s, childbed fever was a common occurrence in European maternity wards, occurring at epidemic levels with a mortality (death) rate of 10%. At two hospitals in Vienna, it was observed patients with doctors providing maternity care were much more likely to die from childbed fever than woman who only had midwives attend to them. In 1846, Hungarian doctor, Ignaz Semmelweis found that the doctors were transferring infections from the patients they were autopsying to the maternity patients in their care, causing the infections. Midwives did not perform autopsies, so were not transferring infections.

A new practice of washing hands with soap and a chlorine solution before examining patients or entering the labour ward reduced mortality within the hospital to <2%

213. The Skin as a First Line of Defence (Page 332)

1 a. *This is the first defence that the pathogens will encounter. Only once the pathogens break past the skin barrier and enter the body do the second (innate) and then third (adaptive) line of defences need to be used.*

b. *Non-specific defence against any type of pathogen. The skin defences will act against any type of pathogen to prevent its entry into the body. There is no developed immunity in the responses.*

2. *The skin provides a physical barrier to prevent pathogens entering the body. Skin secretions (serum and sweat) contain antimicrobial chemicals that inhibit microbial growth.*

3. *Cilia moves the microbes that are trapped in mucus towards the mouth and nostrils, where they can be expelled. Phospholipases kill bacteria by hydrolyzing the phospholipids in cell walls and membranes.*

4 a. *Aids in the maintenance of blood volume (e.g. prevents too much blood loss).*
b. *Prevents bleeding and invasion of bacteria.*

5 a. *Injury exposes collagen fibres to the blood.*
b. *Chemicals make the surrounding platelets sticky.*
c. *Clumping forms an immediate plug of platelets, preventing blood loss.*
d. *A fibrin clot traps red blood cells and reinforces the seal against blood loss.*

6 a. *Clotting factors catalyse the conversion of prothrombin to thrombin, the active enzyme that catalyses the production of fibrin.*
b. *If clotting factors were always present, clotting could not be contained; blood would clot when it shouldn't.*

214. Innate and Adaptive Immune Systems (Page 334)

1 a. *Innate immunity is the system that provides defence against any non-specific pathogen, and destroys pathogens. It is unchanging over time. The adaptive system develops in efficiency over time as the body is exposed to different types of pathogens.*

b. *For pathogens that are encountered for the first time - the immune system is still able to mount a defence to kill the harmful pathogens.*

c. *The immunity improves (and has memory) so that if a pathogen enters the body again, the adaptive system can work much quicker to kill the pathogens - as copies of antibodies are held in reserve.*

215. Phagocytes and Phagocytosis (Page 335)

1. *Chemical markers coat foreign material, marking it as a target for phagocytosis. The markers bind to the phagocyte receptors, triggering engulfment of the foreign material by phagocytes.*

2. *Phagocytosis is the engulfment and removal of a microbe (or other antigenic material) so it cannot harm the organism. Once engulfed*

ISBN: 978-1-99-101423-8

by the phagocyte, the microbe can be safely destroyed by enzymes contained within lysosomes. Phagocytosis is a valuable part of internal defence because it targets all foreign material regardless of whether or not it has been encountered before, so it is a rapid, early protection against infection.

3. *Phagocytes force their cytoplasm out to form a pseudopodium (false foot), which moves them forward in space and wraps around the pathogen to engulf them. Phagocytes can alter their shape to fit between the thin walls of the capillaries and out into interstitial fluid around the cut area to trap pathogens.*

4 a. *Increased diameter and permeability of blood vessels.*

Role: Increases blood flow to the area. Lets defensive substances leak into the tissue spaces.

b. *Phagocyte migration and phagocytosis.*

Role: Phagocytes directly attack and destroy microbes and foreign substances.

c. *Tissue repair.*

Role: Replaces damaged cells and tissues, restoring the integrity of the area.

5. *The site of infection is where the defence process is most active. Damaged tissue, dead phagocytes, and fluid (pus) accumulates as a result of the defensive effort.*

216. Antibodies (Page 337)

1. *B-lymphocytes produce the antibodies as part of an immune response. When specific antigens are encountered then the production of the 'matching' antibody is increased - one type of B-lymphocyte then makes just one type of antibody. The antibody will then bind to and destroy the antigen using a variety of methods.*

2 a. *Agglutinins bind antigens together, inactivating them and stopping them from infecting or damaging cells.*

b. *Antitoxins bind and neutralise toxins, stopping them from damaging cells.*

c. *Chemical markers bind to antigens and act as tags that antibodies recognize. This enhances the ability of phagocytes to engulf and destroy antigens.*

217. Antigens (Page 338)

1. *An antigen is any substance that evokes an immune response in an organism.*

2. *Non-self antigens are foreign substances that the immune system responds to. They originate from outside of the organism (e.g. bacteria or virus). Self antigens are substances that originate from within the body, to which the body's immune system responds.*

3. *Foreign antigens usually mean that a pathogenic (disease causing) organism has entered the body. The body must respond by activating the immune system to ensure its own health is not compromised.*

218. Lymphocytes and Adaptive Immune Response (Page 339)

1. *Bone marrow and foetal liver.*

2 a. *Bone marrow.*

b. *Thymus gland.*

3 a. *Production of antibodies (secreted from B-lymphocytes) against specific antigens. The antibodies disable circulating antigens.*

b. *Involves the T-lymphocytes, which destroy pathogens or their toxins by direct contact or by producing substances that regulate the activity of other immune system cells.*

4. *The presence of an antigen results in the proliferation of specific types of B- and T-cell to target that antigen. First, the antigen is engulfed by a phagocytic dendritic cell, which then presents the antigen on its surface and secretes cytokines. These processes activate T helper cells, leading to the proliferation of various T cell types and the production of antibody-producing B cells.*

5. *Dendritic cells act as messengers by processing antigens and presenting them to T cells so that specific antigens can be recognised and targeted.*

6 a. *Activate T killer cells and other T helper cells. Also needed for B cell activation.*

b. *Destroy target cells on contact (by binding and lysing cells).*

219. Clonal Selection (Page 341)

1. *Millions of B-cells form during development. Each B-cell recognises one antigen only and produces antibodies against it. A pathogen will trigger a response in the B-cell that is specific for it, resulting in proliferation of that B-cell. This is called clonal selection (the antigen selects the B-cell clone that will proliferate).*

2 a. *Plasma cells secrete antibodies against antigens (very rapid rate of antibody production).*

b. *The antibodies produced by a B cell match the specific antigenic receptors on the B cell. When the specific antigen is encountered, the B cell can produce millions of antibodies to the antigen.*

3 a. *Some B cells differentiate into long lived memory cells after the first exposure to an antigen (the cells 'remember' the initial infection).*

b. *They retain a memory of a previously encountered antigen so there is a very quick recognition and response time (to produce antibodies) when they encounter the same antigen again.*

220. HIV/AIDS and the Immune System (Page 342)

1 a. *T helper cells.*

b. *HIV recognizes and binds to specific CD4 receptors on T helper cells.*

c. *Reverse transcriptase transcribes the viral RNA into viral double stranded DNA. The host cell can then transcribe the viral genes.*

2. *The virus population rapidly increases within the first year of infection (initial rapid viral replication), followed by a large drop off in*

ISBN: 978-1-99-101423-8

numbers in the second year (antiviral immune response).

Over years 3-10, the HIV population gradually increases again as the immune system is slowly destroyed.

3 a. HIV infects and eventually destroys the T helper cells of the immune system, reducing the ability of the immune system to fight disease.

b. A person's health will suffer as the immune system's ability to fight disease is compromised. They contract numerous infections and cancers are common.

221. Antibiotics (Page 344)

1. Viruses lack the metabolic machinery, e.g. they have no ribosomes for protein synthesis and membrane transport, that antibiotics can affect.

2. Eukaryote cells do not have the peptidoglycan that antibiotics damage or the specific ribosomes (like bacteria do), therefore antibiotics are relatively harmless to human cells.

3. Bacteriostatic: Chloramphenicol. The bacterial population remains the same over the length of the experiment.

   Bactericidal: Ampicillin. The bacterial population decreases over the length of the experiment.

222. Evolution of Antibiotic Resistance (Page 345)

1. Antibiotic resistance first arises in a bacterial population as a result of mutation in one cell, which is then passed on. If an antibiotic dose is too low to kill all the bacteria with the mutation, some will survive and the resistance will persist. Further mutations that increase resistance will be selected for as antibiotic use becomes more widespread and dose rates increase.

2. If the resistance mutation is in a plasmid, this can be duplicated and exchanged with other bacterial strains (horizontal gene transfer).

3. MRSA has acquired genes for resistance to penicillin and related antibiotics, so these antibiotics are no longer effective against it. This is a problem because S. aureus infections are very common and they are becoming more difficult to treat and control. If resistance continues to spread and susceptibility to different drugs declines, there may be no way of treating these infections in the future.

4. In most cases, bacteria become resistant to an antibiotic within a few years of it first being used. In some cases, resistance is already documented before the antibiotic is released for use. In these cases, the antibiotic would only be used on non-resistant bacteria.

5. Antibiotics cost a lot of money to research and produce. When they are released for use, their useful life is very short and so there is a very short time to recover the costs of development. New technology allows for a quick, cheaper, and effective way to search for alternative antibiotics.

223. Zoonoses (Page 347)

1. Pathogen infects natural host, where it remains and builds up in numbers. Infection of transmission host that comes into contact. Mutates (sometimes), and infects human hosts that consume it or come in close contact with.

2 a. East across the Indian Ocean and the Pacific.

b. The Zika virus has remained around the tropics and subtropics. This is linked to the type of mosquito vector transmitting the disease, because these warmer regions are where the mosquito can live and breed.

3. Zika is transmitted by mosquito bites to humans who live in areas where the mosquito lives, but also through human-to-human contact where bodily fluids are exchanged, such as sexual activity, blood transfusions, and from mother to foetus.

4. Student answers will varying depending on the disease they have chosen. Students may choose to study Covid-19 as an example of a recent pandemic, some details are provided.

   Pathogen: Viral disease caused by the SARS-CoV-2 virus.

   Natural host: A number of suspected hosts, most likely bats, but may also be pangolins.

   Transmission host: Unconfirmed, but most likely sold at the wet market where the disease was first detected.

   Disease originated: Wuhan wet markets, China.

   Disease spread: The first case was notified in December 2019 in China's Hubei Province. The first recorded human-to-human transmission was on 5th January, 2020. By mid January 2020, cases were reported in a number of countries and it rapidly spread around the globe resulting in the WHO declaring a pandemic in March 2020.

   Impact on humans: Covid-19 resulted in millions of deaths and long term health conditions, e.g. long Covid. Medical providers were overwhelmed and people suffering from other health conditions may not have received treatments or surgeries, resulting in unintended health consequences. Many countries went into lockdowns to reduce Covid infections spreading and many social and economic consequences occurred as a result. A vaccine has been developed, but the virus mutates so cannot be eliminated.

224. The Covid-19 Pandemic (Page 349)

1 a. The most likely source (as of 2024) is horseshoe bats.

b. The virus is spread by small droplets from the nose and mouth (e.g. when a person speaks, sneezes, or coughs). People become infected when they breathe these droplets in, or when they touch a surface contaminated with the virus.

c. Viral material was found at the markets. Also the first infections seemed to have happened at specific markets. It was a location that the possible host, transmission species, and humans all came in contact.

2. The more harmful a disease is, the more virulent

ISBN: 978-1-99-101423-8

it is. The more easily a disease transmits from person to person (or animal), the more contagious it is. The Covid-19 disease appeared to be more virulent in the initial variants - and less contagious. The virus has mutated many times to variants that spread easily, but are less harmful.

3 a. All three countries showed exponential growth, however control methods stopped the spread in New Zealand and China. (Smaller numbers in NZ as control methods were implemented when numbers were lower).

b. New Zealand has a lower population and less population density so control methods were able to stop spread faster and at lower numbers of cases. China also implemented strict control measures but the disease had spread more significantly in dense and highly populated areas. USA had a variety of control responses across the states (and increased travel) so the disease continued to spread.

4. Covid-19 disease developed in countries and spread undetected in the earlier period - 'seeding' populations before control measures could be implemented.

5 a.

Region	//	Percentage difference	Percentage change	Percentage difference	Percentage change
		Cases	Cases	Deaths	Deaths
Europe		149%	587%	50.1%	66.8%
Western Pacific		187%	1519%	36.8%	45.2%
Americas		126%	340%	190%	4003%
South-East Asia		2303%	13141%	139%	451%
Eastern Mediterranean		157%	728%	148%	571%
Africa		129%	364%	129%	365%

b. All 6 regions showed an increase; however, South-east Asia showed a significant increase of over 10 x more increases in cases over the year compared to the next highest - Western Pacific.

c. There is more variance in death increases - with the Americas showing a higher proportion of deaths per cases increase.

d. The data highlights the significant of change over the year - esp cases in SE Asia and deaths in the Americas.

6. The death rate per cases, for example Americas, had only 0.21% deaths compared to cases in 2020, while Eastern Mediterranean had 2.46% of deaths compared to cases. The absolute numbers do not show the percentage of the cases that were fatal - and even though the Americas had more cases, the death rate in that initial period of 2020 was much lower. (Note: the deaths were not directly related to the same cases counted).

225. Vaccines and Immunization (Page 352)

1. A vaccine is a preparation of a harmless foreign antigen deliberately introduced into the body to prepare the immune system to resist that pathogen.

2. Vaccines provide protection against dangerous or contagious diseases. They are needed to protect against life threatening childhood diseases (e.g. measles) or to protect against seasonal disease (e.g. flu), to protect against infection after injury (e.g. tetanus), and when travelling to countries with a high incidence of certain diseases (e.g. yellow fever). Antigens in the vaccines enable the matching antibodies to be made and stored by B memory cells.

3. Attenuated viruses are more effective in the long term because they are live (but weakened) and tend to replicate in the body, and the original dose increases over time, so boosters are not needed. This doesn't happen with inactivated viruses, which contain only antigenic fragments (parts of the virus).

4 a. Herd immunity is the protection a non-immunized person gets against a disease when a sufficient proportion of the population has been immunized. Protection is provided because such a large number of the population is immunized that it is harder for a disease to spread.

b. Once the population contains a high proportion of unvaccinated people, herd immunity is lost and a circulating disease can spread very rapidly through the community.

5. Some people have conditions that mean they cannot be vaccinated or they will become ill (e.g. people with immune system disorders or cancer). Herd immunity provides protection to them by reducing the number of potential carriers of a disease and so reduces the chance that they will come into contact with the disease.

6. The preclinical trials were carried out simultaneously, rather than consecutively. Clinical trials were also carried out simultaneously. With large numbers tested, trials were as rigorous as other vaccines, but results were taken at the same time. (Additionally, large numbers of labs were developing the vaccines at the same time so there was more data to be shared around to help all companies).

7 a. To replicate experiments - and ensure the results are the same (and there is no errors in the methods) - this maintains the validity of the results and increases trust in them.

b. If there was an error in the original results (before being peer reviewed) then incorrect information would be given to the public. If these were later to be revised, then the public would lose trust in the science, and potentially lose trust in the ability of the vaccination to work - reducing vaccination uptake.

8. The more dangerous risks from the vaccine were exceeding rare - however the vaccination reduced the chance of catching Covid-19, or if it was caught, then reduced the severity of the symptoms. The danger of developing severe side effects or death from Covid-19 were much greater than those from the vaccination - so

ISBN: 978-1-99-101423-8

overall the vaccination had much less risk to a person - plus it also controlled and slowed the pandemic.

226. Did You Get It? (Page 355)

1 a. Central Nervous System (CNS) and the Peripheral Nervous System (PNS)

b. The automatic system takes place unconsciously - does not require conscious choice from the cerebrum, unlike the voluntary system which requires conscious thought.

c. A sensory neuron begins with receptors on neuron - often on or near the outside of the body, and signals travel to the cerebrum via the spinal cord. The motor neuron originates in the cerebrum (usually) and travels to the effectors, such as muscle or secretary cells.

2. Stress stimulates neurons in the hypothalamus, which sends a signal, via the spinal cord, to the adrenal gland. The adrenal gland releases epinephrine, a hormone. The hormone has various effects on the body to prepare it for instant action, such as increasing heart rate and blood pressure, and converting glycogen to glucose in the liver to supply energy to cells.

3. The contractions of the different layers of muscle around the digestive tract need to be stimulated to coordinate with each other in order for food (bolus) to move effectively - this process needs to occur even when sleeping - so without conscious thought.

4. Stimulus of light causes the auxin to move through auxin efflux carrier in the cells to the shaded tissue side. The auxin increases the transport of hydrogen ions into the apoplast between cells - decreasing the pH and inducing expansins to break the bonds between cellulose so the cells to increase in size - this causes the plants to bend towards the light.

5. A zoonosis disease - as it travels from animal to humans. (The virus transmits from fruit bats to humans).

6. The virus antigens are received by antigen processing cells, such as a dendrite, and these present the antigen to T helper cells. From MHC I pathway the killer cells then converge on the pathogen to destroy it. From MHC II pathway the T helper cells then induce the production of B-lymphocytes, which produce antibodies to kill or neutralize the pathogen - and memory cells keep antibodies for next infection.

Theme C: Interaction and Interdependence

Chapter 12: Ecosystems

227. Features of Populations (Page 358)

1 a. Population growth rate, mortality, birth rate and fertility, age structure.

b. Population growth and birth rates (not declining). Mortality (not high or increasing). Both indicate that the environment can support the population.

2 a. Total abundance, density and distribution, sex ratios, fertility, migration, age structure, sex ratios.

b. Population growth rate, mortality (death rate), natality (birth rate)

3 a. It allows researchers to determine the population attributes of the species and identify reasons for the population decline (e.g. decreased fertility, increased mortality, age or sex imbalance). Strategies to overcome these problems can then be devised.

b. Knowing the population dynamics of fish stocks allows a sustainable harvest to be set so that the stock is not depleted. Growth rate is calculated, taking into account population abundance, and birth and death rates. Harvest is set as a (controllable) source of mortality.

228. Sampling Populations (Page 359)

1 a. E.g. 3, 4, 1, 3, 3, 4

b. Mean dandelions per square: 3

Population estimate: 108

2 a. Mean dandelions per square: E.g 2.67

Population estimate: 96

b. Mean dandelions per square: E.g 3

Population estimate: 108

3. Sampling error: 6 (For all)

4. Based on the examples above, the population lies between 108, and 96 dandelions. We can not know the exact population due to sampling error.

5.

x	$\bar{x}$	$(x - \bar{x})^2$
3	3	0
4	3	1
1	3	4
3	3	0
3	3	0
4	3	1
$\Sigma(x - \bar{x})^2$		6

$\frac{\Sigma(x - \bar{x})^2}{n - 1}$ = 1.2

standard deviation s = 1.09

6. 0.44

7. 1.15 (i.e 3 ± 1.15)

8. 108 ± 41

229. How Do We Sample Ecosystems? (Page 361)

1 a. Quadrat sampling.

b. Belt transect.

c. Area sampling using quadrats.

d. Mark and recapture

2. Aspects of the physical environment can help to explain the distribution and abundance of the species being sampled.

3 a. They might use the school roll and select every 4th person on the school roll.

b. They might divide the school students into males and females or into year groups.

c. They may stop students in the hall between classes to gather information.

ISBN: 978-1-99-101423-8

230. Quadrat Sampling (Page 363)

1. *30 individuals ÷ 37 quadrats = 0.811 per quadrat.*
2. *30 ÷ (37 x 0.08) = 10.1 centipedes per m²*
3. *Probably random although it could be clumped. There is not enough area covered to be sure.*
4. *Presence of suitable microhabitats for cover (e.g. logs, stones, leaf litter) may be scattered.*

231. Sampling a Rocky Shore Community (Page 364)

The results per se are not particularly important, but it is important to understand the method and its limitations. The results will vary depending on a group's agreed criteria for inclusion of organisms in a given quadrat (e.g. when and how an organism is counted when it is partly inside a quadrat). Note: Some algae are almost obscured by organisms or have other algae on top of them.

Population density typical results (total for each category) are:

	A	B	C	D
Plicate barnicle	1667	1296	1852	1667
Oyster borer	370	370	370	370
Snakeskin chiton	185	370	185	185
Coralline algae	3519	2593	2407	3519

Mean no. of organisms per quadrat typical results (total for each category) are:

	A	B	C	D
Plicate barnicle	1.5	1.167	1.67	2.83
Oyster borer	0.33	0.33	0.33	0.167
Snakeskin chiton	0.167	0.33	0.167	0.33
Coralline algae	3.167	2.33	2.167	2.5

1 a. *Population estimate typical results (total for each category) are:*

	A	B	C	D
Plicate barnicle	54	42	60	102
Oyster borer	12	12	12	6
Snakeskin chiton	6	12	6	12
Coralline algae	114	84	78	90

b. *Direct count*

Plicate barnicle	82
Oyster borer	5
Snakeskin chiton	4
Coralline algae	102

232. Estimating Population Size: Quadrats (Page 366)

1. *Results will depend on student investigation area and the population investigated.*

233. Estimating Population Size: Mark and Recapture (Page 367)

1. *Results will vary from group to group. The actual results are not important, but serve as a useful vehicle for discussing the factors that could be altered to improve validity.*

2 a. *(Answers only)*

Species	Population estimate (N = M x n / m)	Biomass (kg)	Biomass by area (kg per ha)
Goldfish	238	86.8	125.7
Koi carp-goldfish hybrids	18	18.4	26.6
Koi carp	158	18.0	26.0
Catfish	116	35.0	50.7
Shortfin eels	540	102.1	147.9

b. *No fish from outside the population can enter and marked fish cannot leave, affecting the results. Thus, the results are likely to be reliable.*

c. *It will tell you what amount of biomass can be supported, in the lake. If the pest fish are removed, then the biomass of shortfin eels could be expected to increase by around the same amount.*

3 a. *Some marked animals may die or leave the area affecting the recapture rate; this would make the results unreliable.*

b. *Not enough time for thorough mixing of marked and unmarked animals.*

4. *Any two in any order:*
   - *Marking does not affect their survival.*
   - *Marked and unmarked animals are captured randomly.*
   - *Marks are not lost.*
   - *The animals are not territorial (must mix back into the population after release).*

5 a. *Territorial or immobile animals.*

b. *These animals will not leave the place they were captured to mix with the population. Recapture at the same location would simply sample the same animals again.*

ISBN: 978-1-99-101423-8

234. Carrying Capacity (Page 369)

1. *Limiting factors are those that limit the population size. A change in the limiting factors may result in an increase or decrease in carrying capacity. For example, a prolonged drought will lower K (water = limiting factor). A sustained increase in food will likely increase K.*

2 a. *Food level has been reduced due to earlier high numbers.*

b. *Available water (and consequently food) are reduced due to the drought.*

c. *Water is more available due to the drought breaking.*

3. *The carrying capacity is dynamic and can change based on changes in the environment. Favourable weather may increase the carrying capacity, but overcrowding may decrease it due to damage to the environment.*

235. Population Regulation (Page 370)

1. *Density dependent factors, such as disease, parasitic infestation, competition, and predation have an increasing effect on population growth as the density of the population increases; their effects are greater at higher population densities because their intensity is determined by the number of organisms present. Density independent factors, such as flood, fire, and drought have a controlling effect on population size and growth that is independent of the population density. The severity of the impact on the population is not correlated with population density.*

2. *When population density is low, individuals are well spaced apart. This can reduce stress between individuals (improving the resistance to diseases) as well as making the transmission of the disease more difficult. Disease is more easily spread between individuals in a crowded population. Epidemics of infectious disease are more likely to occur.*

3. *Feedback results from the interaction of the population and the density dependent factors. High density allows disease to spread easily, reducing the population and population growth leading to low density. Low density slows disease spread. A more healthy population increases to a high density again.*

236. Population Growth (Page 371)

1 a. *Number of individuals dying per unit time*

b. *Number of individuals born per unit time.*

c. *Number of individuals moving into and out of the population per unit time.*

2. *Population growth will be constrained to a level that can be sustained by the factor that is most limiting.*

3 a. $B + I = D + E$

b. $B + I < D + E$

c. $B + I > D + E$

4 a. *14% per year*

b. *2% per year*

c. *20% per year*

d. *−4% per year*

e. *The population is declining.*

237. Population Growth Curves (Page 372)

1.

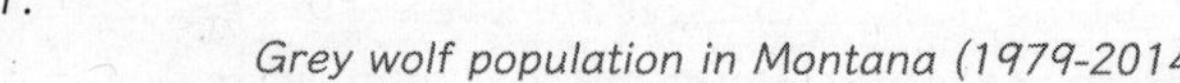

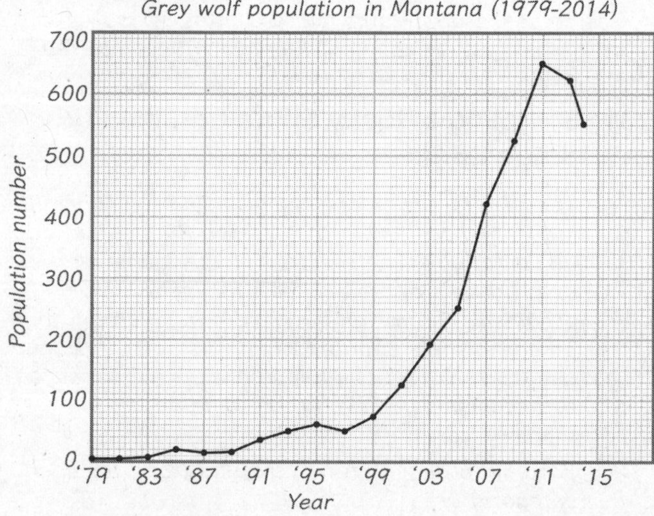

2. *Exponential*

3. *Very few animals were present to add to numbers*

4. *When the wolves moved into the region, there was limited competition for plentiful resources (no other wolves) and the population increased exponentially (with a lag). The population exceeded 600 then declined, indicating a K of around 500-550 wolves.*

5 a. *The growth curve would be a logistic curve.*

b. *The population would grow for a time due to migration and births. Once established, the population will stabilize based on the resources (e.g. jobs, infrastructure, opportunities) and space available (e.g. homes).*

c. *Raid growth at start, then a slight decline, before stabilising around a carrying capacity.*

d. *A straight line up towards the right during growth, then a horizontal line after the population stabilizes.*

e.

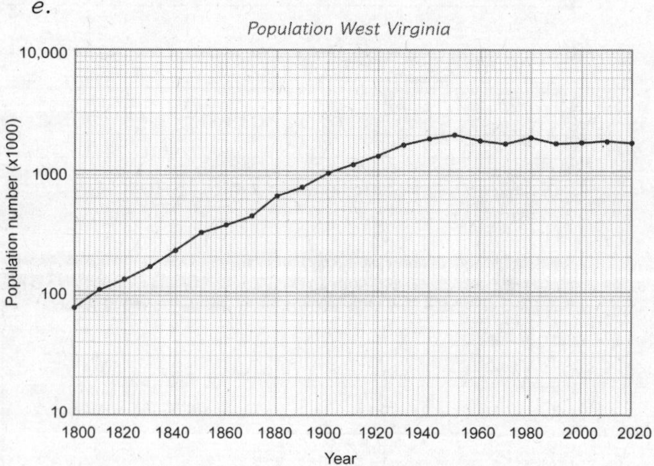

f. *Yes*

ISBN: 978-1-99-101423-8

238. Modelling Population Growth (Page 374)

1. *Students are likely to see exponential growth then a decline . The extent will depend on conditions the yeast is grown in and for how long.*
2. *With space and resources (sugar) the yeast will grow and reproduce rapidly, increasing the population until the resource is used up. Once the resource is used up, the population will decline.*
3. *The population growth may be more sustained but is unlikely to be faster per se as the yeast will have a maximum rate of growth that will likely have been reached.*
4. *The population growth would be sustained (by the addition of new resources) for a longer period of time, although wastes would also build up and affect population growth.*

239. Ecosystems and Communities (Page 375)

1. *A community is a naturally occurring group of organisms living together as an ecological entity. The community is the biological part of the ecosystem. The ecosystem includes all of the organisms (the community) and their physical environment.*

2 a. *Population*

b. *Ecosystem*

c. *The community*

d. *Physical factor*

3. *The resources of an ecosystem are provided by its biotic and abiotic components. When resources are plentiful and diverse, more species can be supported in greater numbers than if resources are limited and of low diversity. Resource limitation tends to increase competition within and between species and reduces species diversity. Diverse, resource rich ecosystems can support a larger number of species interactions (more biotic connections). This, in turn, helps to make processes such as nutrient cycling more efficient.*

240. Interactions Within Species (Page 376)

1. *In scramble competition, individuals compete directly for the same resource and all receive less than they would in the absence of competition. In contest competition, there is a winner and a loser and resources are obtained completely or not at all.*

2 a. *The growth rate of larvae (<u>Rana tigrina</u>) declines as the density increases from 5 to 160 individuals (in the same sized space).*

b. *If there is strong competition for food resources, some larvae (especially those hatching late in the breeding season) will not get enough food to metamorphose and will die. This will limit population size to a level that can be supported by the resources of the environment.*

3 a. *Altruism is behaviour in which an individual disadvantages itself in some way for the benefit of another.*

b. *There is a genetic advantage in helping a related individual. Relatives share some of the same genes so helping a relative to reproduce successfully also increases the chances of some of your own genes being passed on.*

4. *Cooperation reduces the energy an individual expends on essential tasks, leaving more energy for other tasks important to survival. It also increases that task's efficiency, thereby enhancing the survival of the group as a whole.*

5. *Evidence: bird mobbing behaviour (different species act together to drive off a predator). Alone, individuals have poor defence but collectively they can attack a larger predatory bird. When mobbing calls of the black-capped chickadee towards a screech owl are played back, other bird species arrive at the area and engage in coordinated mobbing behaviour.*

241. Interactions Between Species (Page 378)

1. ***Competition***

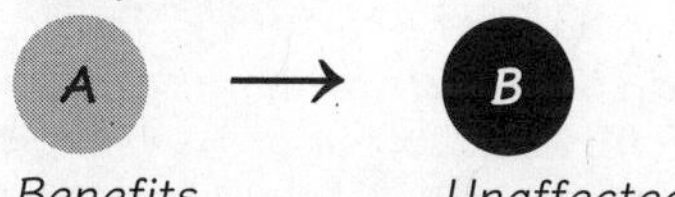

Benefits *Unaffected*

Parasitism

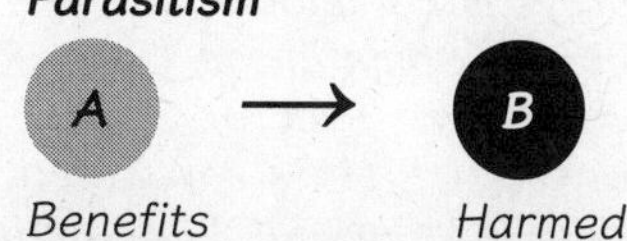

Benefits *Harmed*

Predation/Herbivory

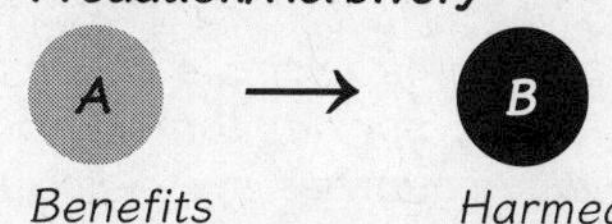

Benefits *Harmed*

Pathogenicity

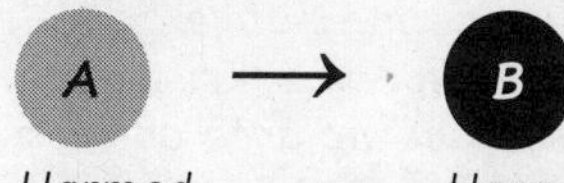

Harmed *Harmed*

2 a. *Competition (Interspecific)*

b. *Each species in harmed (receives less food)*

3. *Parasitism, predation, herbivory, pathogenicity*

4 a. *Mutualism*

b. *The honeyeater benefits from the nectar (food source). The myrtles and proteas benefit from pollination.*

5 a. *Commensalism*

b. *The shrimp benefits from the protection of the anemone. The anemone is unaffected.*

6 a. *Predation*

b. *The dingo benefits, whereas prey is killed.*

7. *Algae photosynthesize to provide coral with energy (glucose/sugars). Coral are animals (need to consume food) but provide algae with a solid structure on which to grow and be protected.*

8. *Both predators and parasite cause harm to the host/prey. Predators kill their prey, parasites normally leave the host alive.*

ISBN: 978-1-99-101423-8

242. Interspecific Competition (Page 380)

1 a. *Interspecific competition refers to competition between individuals of different species for resources. Many examples are possible, e.g. competition between plant species for nutrients, water, or light; competition between scavenging species for a carcass.*

b. *Because individuals of the same species compete for the same resources. In interspecific competition, the competing individuals have some different resource requirements and this reduces the intensity of the competition.*

c. *Interspecific competition is generally less effective than intraspecific competition at limiting population size because different species will generally switch to alternative resources when contested resources are depleted. As long as alternatives are available, population size won't be resource limited.*

2 a. *The two species have similar niche requirements (similar habitats and foods). Reds once occupied a much larger range than currently. This range has contracted steadily since the introduction of the greys, which points to the greys being the cause of the declines.*

b. *The evidence is not conclusive. The introduction and increase in greys is correlated with a decrease in reds but in some areas the species coexist and in other areas (such as Scotland) reds predominate. Other factors implicated in the decline include loss of suitable habitat (particularly coniferous forest) and spread of disease.*

3. *Features will depend on the example chosen. Typical examples for two species that are proving to be invasive in many countries are described:*

Mallard duck (Anas platyrhynchos)

The biology and behaviour of Mallard ducks result in them out competing native or endemic species. If the endemic species has a narrow niche, it may be lost from the area or even become extinct.

Features of the mallard duck include:

- *Clutch size is relatively large compared with many other duck species.*
- *Mallards are sexually aggressive and are capable of interbreeding with several native duck species where their ranges come to overlap (gray ducks in New Zealand, American black duck, the Florida mottled duck, and the endangered Hawaiian duck).*
- *Mallards bully native duck species in competition for food and nest sites.*
- *Mallards are adaptable in different environments.*

Mosquito fish (Gambusia affinis):

Mosquito fish are aggressive, breed quickly and will attack other fish species and prey on the eggs and larvae of native fish and amphibians,and cause a decline in biodiversity. The tolerate a wide range of water temperatures and are well adapted to out-compete other species.

Traits of the mosquito fish include

- *Prolific breeders in many diverse environments.*
- *Aggressive competitors for food.*
- *Generalist feeders, and will prey on the eggs and larvae of native fish and frogs.*
- *Reproductive strategies tuned to maximizing reproductive output. Females are able to store sperm for extended periods and colonize environments without needing to meet a mate there.*

243. Testing for Interspecific Competition (Page 328)

1 a. *Resource competition can exclude one species if the superior competitor is able to deplete resources to the point that the resources cannot be accessed by the less able competitor. The diatom Synedra can extract silicates to a lower level than Asterionella can, so when they are grown together Asterionella cannot obtain enough silica to build its cell walls. Asterionella then declines to (experimental) extinction.*

b. *Natural environments have many different factors contributing to the outcome of competitive interactions, e.g. predators, changes in temperature, microhabitats, alternative resources etc. These factors are not completely represented in laboratory environments. Because of this, the actual outcome of competition in natural communities may be very different.*

2. *In this case, Asterionella outcompetes Synedra. This may be because the cellular processes of Asterionella are more efficient at 8°C than are Synedra's.*

3. *Sugar gliders are mostly found in rainforests whereas squirrel gliders are found in other forest types although there is some overlap.*

4. *The gliders occupy different environments. Sugar gliders move to rainforest while squirrel gliders occupy drier forests.*

5. *The sugar glider can live in both rainforest and dry forest environment. Where the species have range overlap, the sugar glider restricts itself to the rainforest, presumably to avoid competition.*

6. *In testing a hypothesis by experiment, conditions can be controlled and many tests can be run simultaneously and repeated to gather accurate data. When observing, data can only be gathered on what can be seen, along with any historic data already gathered. It may take a long time to gather data and it relies on the amount of sampling. More interpretation is required and many uncontrollable factors may be involved.*

244. Using the Chi-Squared Test in Ecology (Page 384)

245. Chi-Squared Exercise in Ecology (Page 385)

1. *There is no association between lesser pond sedge and marsh bedstraw.*

2. *LPS and MBS were found together in 11% of*

ISBN: 978-1-99-101423-8

samples. When LPS was counted present, MBS was present less often than it was absent. When LPS was counted absent MBS was absent more often than it was present.

3.

	LPS present (1)	LPS absent (0)	Total
MBS present (1)	5.88	8.12	14
MBS absent (0)	36.12	49.88	86
Total	42	58	100

4.

Category	O	E	O–E	$(O-E)^2$	$\frac{(O-E)^2}{E}$
LPS 1/MBS 1	11	5.88	5.12	26.2	4.45
LPS 0/MBS 1	3	8.12	-5.12	26.2	3.2
LPS 1/MBS 0	31	36.1	-5.12	26.2	0.73
LPS 0/MBS 0	55	49.8	5.12	26.2	0.53
Total = 100				χ^2 = 8.91	

5. 1
6. 0.005
7. Reject
8. LPS and MBS are found associated more often than you would expect by chance alone.

246. Predator-Prey Relationships (Page 386)

1 a. Food for the hare fluctuates seasonally (less in winter), creating cycles of abundance (higher and lower numbers). The lynx rely on the hare for food, so their population numbers fluctuate in the same way.

b. It takes time for the lynx numbers to respond because they must gain more energy and then produce and raise more young.

c. The snowshoe hare is almost the only species the lynx eats, so it relies heavily on it for food and has few chances to switch to other prey when hare numbers are low. Therefore, when hares numbers fall, lynx numbers will also fall.

2. These birds move around to follow prey populations and so may move into and leave areas as quickly as prey populations increase and decrease. Thus, their numbers increase or decrease more quickly than less mobile populations in which only births increase their numbers.

3 a. The number of lemmings eaten by predators increases rapidly as the lemming population density increases.

b. Long tailed Skua

c. The birds are likely very mobile and feed on many other species so, although it takes advantage of an increase in lemming populations, it is not dependent on them to maintain its population.

d. The predator populations may fluctuate less than they currently do. Predator numbers may not fall as low previous low years, and also will not rise as high. They may, over time, decline to low numbers depending on lemming populations.

247. Control of Populations (Page 388)

1 a. The Guanay population declines rapidly as the tonnage of anchovies caught rapidly increases.

b. Bottom-up control. The anchovies are the guanay's prey. The fishing of anchovies leaves no food for the auanay and so their population declines.

2 a. Top-down control. The sea otters eat the sea urchins and control their population. The sea urchins eat the kelp and cause large barren areas of sea floor.

b. The kelp would likely regrow as the sea urchins would be controlled by the sea otters.

248. Chemical Defences (Page 389)

1 a. Allelochemicals are chemicals that are produced by plants that affect the growth or germination of competing plants.

b. Allelochemicals could be used as herbicides because they prevent the germination of other plant seedlings or kill plants that come in contact with them.

2. Plants and fungi produce antibiotics to reduce competition (fungi) and prevent disease (in plants).

249. Systems and Energy (Page 390)

1 a. Open

b. Isolated (ideally)

c. Closed

d. Open

2. Ten percent law (thermodynamics)

3. A law is a description that predicts an outcome based on certain conditions. A theory explains why a phenomenon occurs. E.g. Newton's law of universal gravitation, and the theory of gravity, which explains that law.

250. Flow of Energy Through Ecosystems (Page 391)

1. Photoautotrophs and chemautotrophs are able to manufacture their own food from simple inorganic substances (e.g. CO_2) using either the free energy in sunlight or chemical energy. Heterotrophs cannot manufacture their own food. Their energy is obtained by consuming (feeding) on other organisms. Thus, energy flows from autotrophs to heterotrophs and then to decomposers and back into the environment

2 a. Wastes may contain energy rich molecules, e.g. urea, and undigested matter, e.g. cellulose.

b. Some energy from the breakdown of glucose is lost as heat energy to the environment.

3. Most of the energy available to each trophic level is lost in respiration; wastes, death, or consumption.

4. Chemoautotrophs are able to fix carbon to form organic molecules, filling the primary producer role in the deep sea environment. They provide the first trophic level for deep sea ecosystems, which do not receive light from the sun.

5. Although chemoautotrophic bacteria can use chemical energy to fix carbon, many do

ISBN: 978-1-99-101423-8

this aerobically and require oxygen (as a final electron acceptor) which is provided by photosynthetic organisms in the surface waters of the oceans. The animals that live off the bacteria also require oxygen for respiration.

251. Quantifying Energy Flow in an Ecosystem (Page 393)

1 a. 14,000
b. 180
c. 35
d. 100
2. Solar energy (Sun)
3 a. Photosynthesis
b. Eating/feeding/ingestion
c. Respiration
d. Export (lost from this ecosystem to another)
e. Decomposers and detritivores feeding on same.
f. Radiation of heat to the atmosphere
g. Excretion/egestion/death
4 a. 1,700,000 ÷ 7,000,000 x 100 = 24.28%
b. It is reflected.
5 a. 87,400 ÷ 1,700,000 x 100 = 5.14%
b. 1,700 000 - 87,400 = 1,612,600 (94.86%)
c. Most of the energy absorbed by producers is not used in photosynthesis. This energy, which is not fixed, is lost as heat (the heat loss component before the producer level is not usually shown on energy flows).
6 a. 78,835 kJ
b. 78,835 ÷ 1,700,000 x 100 = 4.64%
7 a. Decomposers and detritivores
b. Transport by wind or water to another ecosystem (e.g. blown or carried in water currents).
8 a. Low oxygen conditions (hypoxic) or lack of oxygen (anoxic), low temperature, low moisture.
b. Energy remains locked up in detritus and is not released.
c. Geological reservoir
d. Fossil fuels such as oil, natural gas, and coal. Oil and natural gas are formed from the remains of marine plankton. Coal and peat are both of plant origin. Peat is partly decomposed and coal is fossilised.
9 a. 87,400 → 14,000: 14,000 ÷ 87,400 x 100 = 16%
b. 14,000 → 1,600: 1,600 ÷14,000 x 100 = 11.4%
c. 1,600 → 90: 90 ÷ 1,600 x 100 = 5.6%
d. Producers to primary consumers

252. Food Chains (Page 395)

1 a. Producers: obtain energy directly from the sun via photosynthesis.
b. Consumers: obtain energy by eating other organisms
c. Detritivores: obtain energy from eating dead organic matter.
d. Decomposers: obtain energy by extracellular digestion of dead material.
2 a. (see diagram below)

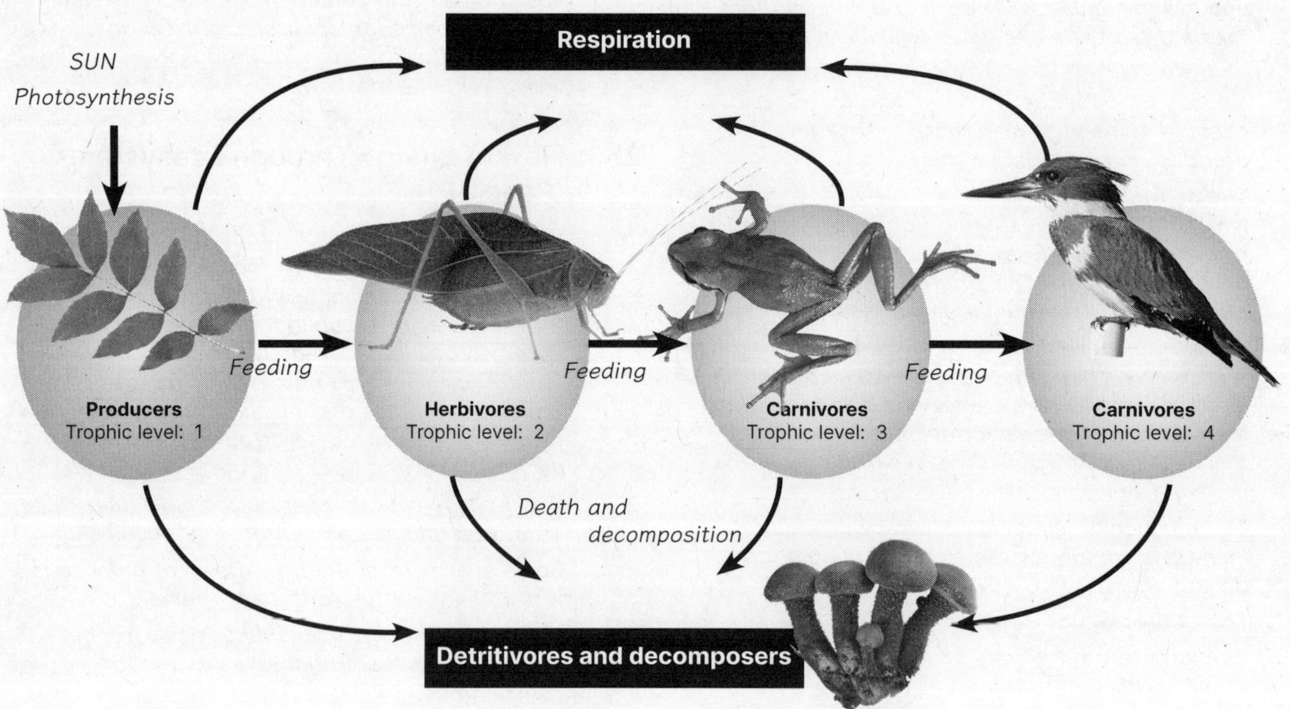

b. The Sun

ISBN: 978-1-99-101423-8

253. Food Webs (Page 396)

1. *Energy passes through ecosystems by being transferred from producers to consumers. Some energy is lost at each transition in metabolic wastes and as heat.*

2 a. *The amount of energy available to each successive trophic level decreases along the food chain.*

b. *Energy is lost as heat and as organic waste at each trophic level. This energy is not available to the next trophic level. Heat energy cannot be used by organisms although the energy in organic waste is available to detritivores and decomposers.*

3. *In a grazing food web, energy moves from producers to primary consumers and then to secondary consumers. All consumer groups provide energy to decomposer levels. In a detrital food web, producers provide energy as dead plant material, and the primary consumers are decomposers. Energy flows back and forth between decomposers and detritivores but herbivores and carnivores do not feature.*

4 a. *Autotrophic protists → Daphnia → Diving beetle*

b. *Autotrophic protists → Daphnia → stickleback → pike*

c. *Macrophyte → great pond snail → herbivorous water beetle → stickleback → pike*

d. *Macrophyte → carp → pike*

e. *Autotrophic protists→ mosquito larva → Hydra → dragonfly larva → carp → pike*

f. *Macrophyte → herbivorous water beetle → carp → pike*

g. *Autotrophic protists → Daphnia → Asplanchna → leech → dragonfly larva → carp → pike*

h. *Detritus → Paramecium → Asplanchna → leech → dragonfly larva → carp → pike*

i. *Detritus → Paramecium → Asplanchna → leech → dragonfly larva → diving beetle*

j. *Detritus → great pond snail → leech → dragonfly larva → carp → pike*

k. *Detritus → great pond snail → leech → dragonfly larva → diving beetle*

l. *Detritus → Paramecium → mosquito larva → Hydra → dragonfly larva → carp → pike*

5. *(See diagram below)*

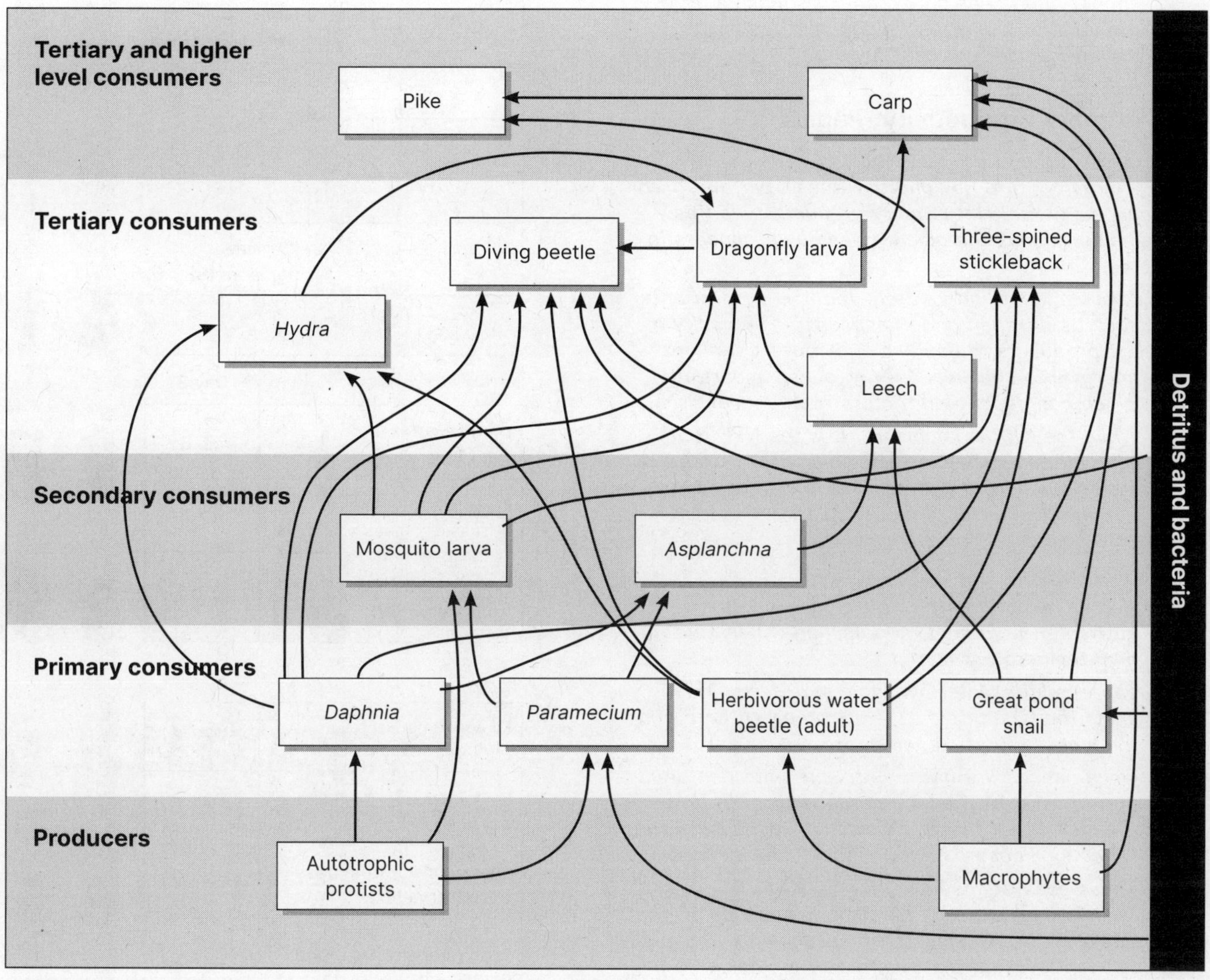

ISBN: 978-1-99-101423-8

254. Ecological Pyramids (Page 399)

1 a.

	Energy in organisms kJ m-2 yr-1	*Energy lost (respiration/ waste)*	*Energy passed to next level*	*% available to next level*
Producers (P)	*31,897*	*27,279*	*4618*	*14%*
Herbivores (C1)	*4618*	*4154*	*464*	*11%*
Carnivores (C2)	*464*	*444*	*20*	*4.5%*
Top carnivores (C3)	*20*	*20*		*0%*

b.

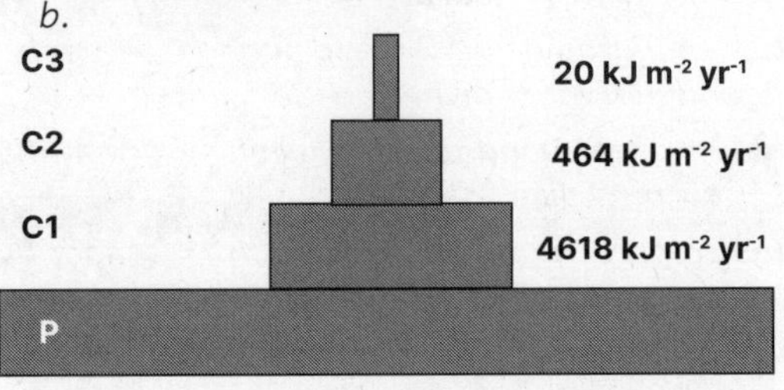

c. *To invert an energy pyramid would require more energy to be available to each level above. This cannot happen as only producers make energy available to higher levels and energy is lost to the environment at each level.*

255. Primary Productivity (Page 400)

1 a. *Gross primary productivity is the amount of energy captured by photosynthesis per unit area per unit time. Net primary productivity is this amount minus the amount used by producers in respiration.*

b. *Availability of light, water, and limiting nutrients such as nitrogen and phosphorus. The ability of the producers to capture and convert sunlight energy into chemical energy is also important, which can depend on factors such as leaf area.*

c. *While the GPP is important, it is the NPP that will determine the consumer biomass that can be supported by the system. This is important for agricultural systems in particular.*

2 a. *Estuaries, swamps and marshes, tropical rainforest*

b. *Non-limiting supplies of nitrogen (and other nutrients), high light, non-limiting water, and high temperatures.*

c. *There is little water available in deserts. This limits the type of plants that can grow there and the biomass that can be supported.*

d. *Open oceans are low in nutrients (nitrogen and phosphorus) needed for producer growth and the lack of light limits productivity because water reflects or absorbs a lot of the incoming light.*

3. *These ecosystems have high light levels, unlimited water, and non-limiting supplies of nutrients (from release from sediments, internal nutrient cycling, and freshwater inputs, including from land runoff).*

256. Trophic Efficiencies and Secondary Production (Page 402)

1. *Table 1*

Day 1	**Day 3**	
30 g	11 g	g consumed = *19 g*
–	2.2 g	
0.2	*0.2*	
109.20	*40.04*	kJ consumed per 10 larvae = *69.16*
10.92	*4.004*	kJ consumed per larva **(E)** = *6.92*

Table 2

Day 1	**Day 3**	
0.3 g	1.8 g	g gained = 1.5 g
0.03	*0.18*	g gained per larva = *0.15*
–	0.27 g	
0.15	*0.15*	
0.10	*0.62*	kJ gained per larva **(S)** = *0.52*

Table 3

	Day 3
Dry mass frass from 10 larvae	0.5 g
Frass energy (waste) = frass dry mass x 19.87 kJ	*9.94*
Energy from frass from 1 larva **(W)**	*0.99*

Table 4

kJ consumed per larva **(E)** =	*6.92 kJ*
kJ gained per larva **(S)** =	*0.52 kJ*
Energy (in kJ) from frass from 1 larva **(W)** =	*0.99 kJ*
Respiratory losses (in kJ) per larva =	*5.41 kJ*

2 a. *0.52kJ*

b. *(S ÷ E) x 100 = (0.52 ÷ 6.92) x 100 = 7.47%*

ISBN: 978-1-99-101423-8

c. *Yes, it is less than 10%, but is a realistic value for efficiency of transfer from producers to primary consumers.*

3 a. *0.225 g*

b. *0.225 ÷ 10 ÷ 3 = 0.0075 g d^{-1}*

257. The Carbon Cycle (Page 403)

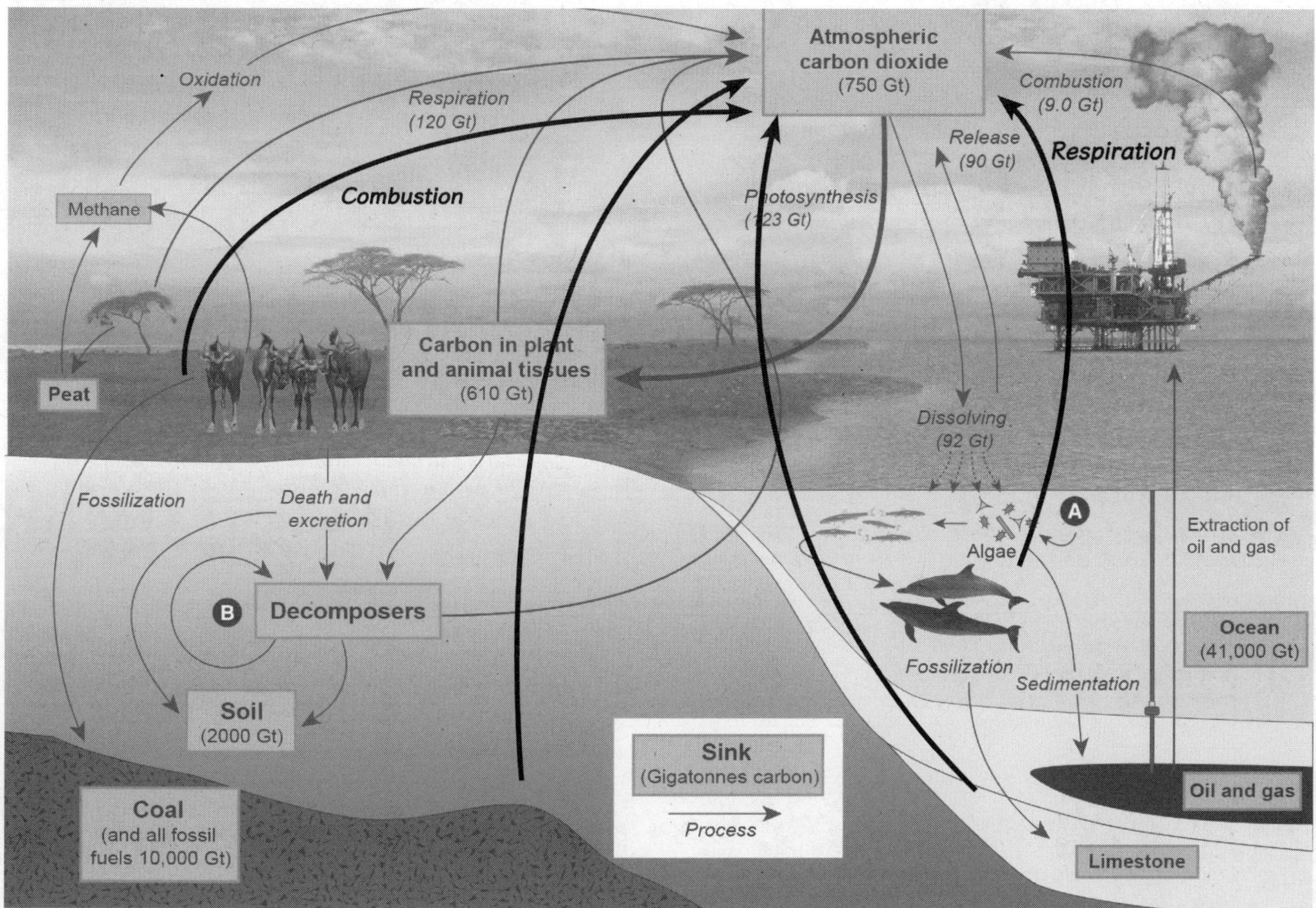

2 a. *Respiration (stepwise oxidation of glucose) and combustion (rapid oxidation of organic substances accompanied by heat).*

b. *Both involve the release of carbon dioxide.*

3 a. *Atmosphere*

b. *Coal*

c. *Limestone*

d. *Oil and natural gas*

4 a. *Photosynthesis*

b. *Respiration*

5. *Carbon would eventually be locked up in the bodies (remains) of dead organisms. Dead matter would not rot. Possible gradual loss of CO_2 from the atmosphere.*

6 a. *Plant material trapped and compressed under sediment in swampy conditions.*

b. *Algae and zooplankton (animal plankton) buried and compressed under sediment.*

c. *Compressed accumulated remains of marine organisms with calcium carbonate skeletons.*

d. *Partly decayed vegetation formed because of acidic or anaerobic conditions.*

7. *Some carbon may spend only a short time in a reservoir. Carbon generally cycles rapidly through living organisms as they live, metabolize, and then die, although some large, long-lived organisms, such as large trees can lock up carbon for tens or even hundreds of years. Carbon may stay much longer in geological reservoirs such as limestone and fossil fuels (tens to hundreds of millions of years for coal).*

8. *Living organisms are a link between the atmosphere and the geosphere. Plants take in carbon from the atmosphere. If they die and are buried without decomposition, their carbon is transferred to the geosphere (e.g. as coal). Animals and decomposers return the carbon in plants to the atmosphere via respiration. Fossilization can trap carbon in limestone.*

9. *Human activity returns the carbon to the cycle at a much greater rate than natural processes. It also returns it mostly as carbon dioxide in the atmosphere.*

10 a. *Increasing the rate of photosynthesis will remove CO_2 from the atmosphere and store it in organic matter (wood).*

b. *Increasing the rate of respiration will increase the amount of CO_2 in the atmosphere.*

c. *Combustion*

11. *Nutrient cycling ensures nutrients are available to all organisms. Without nutrient cycling, nutrients would become permanently locked away in sinks and organisms would suffer deficiencies, ultimately leading to extinction.*

ISBN: 978-1-99-101423-8

258. Analyzing Changes in Atmospheric Carbon Dioxide (Page 406)

1. *Approx 115 ppm*
2. *Approx 6 ppm*
3. *Over winter (Dec-Feb), respiration increases compared to photosynthesis (which is at a low) so CO_2 peaks. This reverses in late spring (May). During summer/early autumn/fall (Jul-Sep), photosynthesis increases and CO_2 falls again.*
4. *Midnight is the lowest point of photosynthesis and the highest point of respiration, thus atmospheric CO_2 peaks around midnight*
5. *The atmosphere would progressively warm over time.*

---

259. Did You Get It? (Page 407)

1 a. *Parasitism*
 b. *Cat is harmed. Tick benefits.*
 c. *The tick population would decline (detrimental effect) as there would be fewer hosts available to support them.*

2. *Population fluctuations depend on environmental factors such as the amount of food available, nesting space, or the change of seasons etc. When conditions are favourable, a population will enter exponential growth until resources are used up or predators are able to suppress the growth. The population then crashes until conditions become favourable once again.*

3 a. *A*
 b. *D*
 c. *10,000 J (1,000,000 ÷ 10 ÷ 10)*

4. *Decomposers fit outside the food chain because they derive energy from each level as each level produces waste or dies and decomposes.*

5 a. *Primary production decreases with depth. Light levels decline with depth and close to 100 m are too low to support much photosynthesis.*
 b. *Photosynthesis occurs in the photic zone. Producers in the photic zone provide the basis of the marine food chain (most of the primary and higher order consumers), so most marine life is found here.*

---

260. Summary Assessment (Page 408)

1. *c*
2. *a*
3. *d*
4. *a*
5. *c*
6. *d*
7. *d*
8. *b*
9. *Shoots bend towards the light because auxin causes cells on the shadowed side to lengthen. The opposite occurs in the roots and so they bend downwards (in the opposite direction - angle).*
10. *An action potential triggers the influx of calcium in the presynaptic cells, which induce vesicles to release acetylcholine (ACh). ACH diffuses across the membrane to the postsynaptic cell, attaching to the ACh receptor, which opens and causes an influx in sodium ions in the postsynaptic cell and triggers another action potential.*

11 a. *Cell A is an antigen-presenting cell, e.g. dendritic cell.*
Its role is to present antigens to T helper cells.
 b. *B represents chemical molecules called cytokines. In this particular part of the immune response, their role is to stimulate the production of T helper cells so the number of T helper cells increase.*
 c. *T helper cell*
 d. *B cell, specifically a plasma B cell.*
 e. *E is an antibody, a protein secreted by plasma B cells. Antibodies bind and destroy a specific antigen.*

12. *Aerobic respiration uses oxygen as an electron acceptor at the end of the electron transport chain. Anaerobic respiration can involve several different pathways including glycolysis, lactic and alcoholic fermentation, and non oxygenic respiration involving an electron transport chain.*

13 a. *Competition*
 b. *Interspecific*
 c. *Both are harmed*
 d. *Competition for food/space (between and within animal species), Competition for light/nutrients/ space (between and within plant species).*

14. *Photosynthesis produces oxygen and stores sunlight energy as chemical energy (in glucose). It removes carbon dioxide from the atmosphere. Respiration releases the energy stored in glucose, removes oxygen from and releases carbon dioxide into the atmosphere.*

15. *Temporal summation: several nerve signals from a single axon arrive in quick succession and sum to produce an action potential. Spatial summation: signals arrive from different axons and sum to produce an action potential.*

Theme D: Continuity and Change

Chapter 13: Molecules

261. DNA Replication (Page 413)

1. *The stages have been provided as numbered steps for conciseness.*
 1) *DNA helix unwinds – Helicase enzyme breaks the hydrogen bonds and separates strands at replication fork.*
 2) *Primase catalyses short RNA primers to begin replication.*
 3) *DNA polymerase (III) attaches to DNA at RNA primer sequence – moves from 5' to 3' direction on leading strand and 3' to 5' on lagging strand. DNA polymerase I processes RNA primers to remove from final replicated strands.*
 4) *DNA polymerase (III) enables the connection of new nucleotides (using base pairing rules A-T and C-G) to form the replicated strand.*

ISBN: 978-1-99-101423-8

2. The results show that DNA replication is semi-conservative. Generation 0 occurred before any replication has happened, therefore all the DNA is the original heavy. The 1st generation has intermediate (one new, one original strand), and the 2nd generation has only one DNA strand that is original, and 3 that are newly formed.
3. DNA replication prepares a chromosome for cell division by producing two chromatids which are identical copies of the genetic information for the chromosome. DNA replication is required for reproduction (called binary fission in bacteria), for growth and repair of cells - where identical replacement cells are required in multicellular organisms.
4. There would not be enough genetic material to produce two cells with the genetic information necessary for life's functions. DNA replication creates genetic material from 'building blocks' of amino acids found in the cell.
5. New cells for growth need to be identical to those they are replacing, i.e. an organism contains the same DNA in each of its cells. Non-identical cells, such as those with mutations, can cause failure of organs/body.

262. Details of DNA Replication (Prokaryote) (Page 415)

1. Enzymes catalyse the reactions that occur during DNA replication. They unwind the DNA, copy the DNA strands, rejoin DNA sequences, and proofread the DNA to correct mistakes. Enzymes are important in ensuring the same sequence of DNA in the parent cell occurs in the daughter cells (fidelity).

2 a. Unwinds the double strand parent DNA to allow replication.

b. Synthesizes an RNA primer to indicate the start position of replication on a strand.

c. Digests the RNA primer when no longer required - and replaces with DNA - so the new strand matches original template strand.

d. Adds nucleotides from the RNA primer onwards in the 5' to 3' direction (leading strand and Okazaki fragments).

e. Joins Okazaki fragments into a continuous length of DNA.

3. The strand end with a phosphate group on the carbon 3 is also called the 3' end. The strand end with a phosphate group on the carbon 5 is also called the 5' end. DNA synthesis is in the direction of 3' to 5'
4. On the leading strand the DNA replication is continuous from the RNA primer towards the Helicase opened end, so 3' to 5' on the leading strand. On the lagging strand the DNA replication still needs to occur 3' to 5', but as this in the opposite direction, only short segments can be made - with repeated RNA primer sections. These segments are called Okazaki fragments, and are joined into a continuous strand by DNA ligase.
5. Any mutations (incorrect base pairing) created during replication are located and corrected before protein synthesis (as most mutations cause non-functional products). DNA polymerase (I, II, and II) can proofread consecutively as DNA is replicated. A removal enzyme (exonuclease) is used to remove the wrong nucleotide and allows a new, correct nucleotide to be added instead.

263. Polymerase Chain Reaction (Page 417)

1. The material needs to be heated to a temperature that would make the eukaryotic polymerase unable to function (i.e. it would denature it), so therefore Taq polymerase is used from organisms (heat tolerant bacteria/archaea) that is still able to function despite the high temperatures due to its adaptation to the high temperatures.
2. The ability of PCR to amplify and clone small amounts of DNA means that, even if just a few molecules are present (e.g. on ancient bones, or an old blood stain), the DNA may be able to be extracted and amplified into millions of molecules. This allows research such as identifying a species or individual to be done. In bioengineering, PCR means a DNA molecule that may have been isolated from an organism or made synthetically can be copied many times. The copies can be used for further research, such as insertion into subjects (e.g. mice or bacteria) for research or production or proteins.
3. The mixture of DNA is heated to a high temperature so that the DNA is denatured and the strands can be separated before the DNA primers are added.

4 a. 1024

b. 33,554,432 (33.5 million)

264. Gel Electrophoresis (Page 418)

1. The purpose of gel electrophoresis is to separate mixtures of molecules (proteins, nucleic acids) on the basis of size and other physical properties. The similar particles can then be analysed in comparison to other samples - i.e. profiling.

2 a. The frictional force of each fragment's size (larger fragments move more slowly than smaller ones).

b. The strength of the electric field (movement is more rapid in a stronger field).

3. The gel is full of pores (holes) through which the DNA fragments must pass. Smaller fragments pass through these pores more easily (with less resistance and therefore faster) than larger fragments.

265. Applications of DNA Tools (Page 419)

1. Profiles of everyone involved must be completed to compare their DNA to any DNA found at the scene and therefore eliminate (or implicate) them as suspects.
2. The alleged offender is not guilty. The alleged offender's DNA profile does not appear in the DNA collected at the crime scene nor does it appear in the DNA database. Profile E's DNA is found at the scene.
3. 14-10, 14-11, 15-10, 15-11

ISBN: 978-1-99-101423-8

4 a. No

b. The man cannot be the biological father because there are two mismatches in the profiles. The child does not show any matches with STR D19S433 and D2S441.

5 a. Each whale species has a distinct DNA profile. Profiling the whale meat therefore reveals the types of whales they meat came from.

b. Individual whales also have their own DNA profile. Profiling the whale meat from each species reveals how many whales of that species were killed (by simply counting the number of different profiles.

266. What is Gene Expression? (Page 421)

1. A gene is a section of DNA that codes for a protein (or other mRNA product).

2 a. Gene expression is the process of decoding a gene to make a protein.

b. i. Transcription: The DNA is rewritten into a primary RNA transcript.

ii. Editing: Introns are removed from the 1° transcript.

iii. Exons spliced together to form mature mRNA.

3 a. Multiple copies of genes are produced.

b. A large amount of protein is produced which is useful for fast growing larval stages.

267. The Genetic Code (Page 423)

1.

ATG	GGT	TAC	CTG	AGG	GTA	ATA	CGG	GCA	CTT	TAG

1 a & b.

mRNA

AUG	GGU	UAC	CUG	AGG	GUA	AUA	CGG	GCA	CUU	UAG

Ameno acids

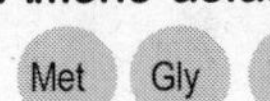

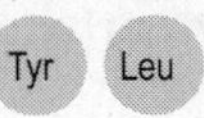

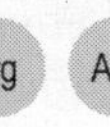

2. They are the same except for U replaces T in mRNA.

3. The degeneracy of the code guards against the possible detrimental effects of single base changes (mutations) to the code because several different codons encode the same amino acid. There is a good chance therefore that a point mutation will not cause a change to the amino acid sequence of the protein encoded.

4. Although different codons may encode the same amino acid (redundancy), no one codon encodes more than one amino acid so the code is not ambiguous. This is important because minor changes may not affect the protein (due to redundancy) but any particular codon can only mean one amino acid, so the coded message is clear.

5 a. GUU, GUC, GUA, GUG

b. GAU, GAC

6 a. 6 ways

b. 4 ways

c. Methionine (Met), tryptophan (Trp)

268. Transcription in Eukaryotes (Page 424)

1 a. RNA polymerase

b. The template strand

c. complementary to

d. Uracil

2 a. 5' to 3'

b. RNA polymerase adds nucleotides to the 3' (-OH) end of the growing RNA thus the RNA transcript extends from the 3' end of the chain.

3 a. It is always where translation will begin.

b. ATG

269. Translation (Page 425)

1. The ribosome is the organelle that the mRNA attaches to so translation can occur. The mRNA binds to the small subunit and the tRNA with their amino acids arrive in the A site of the large ribosomal subunit. The mRNA carries a complementary strand that has been transcribed from the original DNA. A triplet of mRNA nucleotides form a codon. The matching (complementary) tRNA will attach to the codon of the mRNA at the ribosome, with a specific amino acid attached. Beginning with the start codon (amino acid MET) each consecutive amino acid will attach with a peptide bond to form a peptide chain. The tRNA will leave and be recycled.

2. tRNA molecules transport amino acids to ribosomes where the anticodons of the tRNAs are paired with the codons on the mRNA. tRNA enters the ribosome at the A site. Amino acids join together to form a polypeptide chain at the P site. Unloaded tRNAs are released from the ribosome at the E site.

3. The production of many polypeptide chains at the same time. This is important for other processes and reactions that depend on those polypeptides being available in sufficient quantity.

4. The first/start amino acid (MET) arrives in the ribosome and the next tRNA delivers the second amino acid. The amine end (NH_2) of one amino acid reacts with the carboxyl end (COOH) of another. A hydrolysis reaction removes 2 x H and 1 x O (H_2O) and the peptide bond forms between the C and N. Once the peptide chain reaches the stop codon/amino acids, it breaks away and moves out from the ribosome to be further modified.

5. For example: A recent discovery of a deletion mutation, a type of point mutation, called CCR5-delta 32, in some Eurasian populations (nearly 14% of the population have the mutation) creates a change in the protein that forms a receptor on the surface of T cells. The HIV virus cannot bind to this receptor and therefore will not cause an infection. The resistance is highest in homozygous carriers (both mutated genes), but heterozygous carriers (one gene) also show resistance to HIV infection.

ISBN: 978-1-99-101423-8

270. Regulating Transcription (Page 427)

1. *Transcribed but untranslated regions have a role regulating the expression of the gene or export of the mRNA from the nucleus. The untranslated region forms a structure that attaches to transcription factors and provides a site for RNA synthesis to begin translation.*

2 a. *A transcription factor is a protein with a role in creating an initiation complex for RNA Polymerase II binding and transcription.*

b. *Regulatory genes.*

c. *Activator proteins attach to the enhancer end of the RNA. This bends towards the promoter region with transcription factors in between to form a loop. The RNA polymerase II then attaches to the promoter region to begin translation.*

3. *The cap protects the mRNA from degradation. The tail aids export, translation, and stability.*

4 a. *They are removed from the 1° transcript.*

b. *Introns may be processed to produce regulatory elements.*

5. *Primary mRNA can be spliced and combined in many different ways to produce many protein variants. Proteins are also modified after translation.*

6. *Each gene produces 40 proteins on average (1,000,000÷25,000).*

271. Regulating Translation (Page 429)

1. *Connect the initiator anticodon to the small ribosome subunit, and then attach to mRNA so it can locate the start codon, and where it can begin translation.*

2. *The P site is where tRNA initially start translation - the next tRNA arrives at the A site, to the right, where polypeptide bonds form. 'Empty' tRNA move to the E site, then leave, the middle tRNA moves to the P site, then the process repeats.*

3. *(1) Removal of signal sequences (pre-proinsulin). (2) Adding disulfide binds and forming proinsulin. (3) Cleaving the chain to form smaller molecules, removal of connecting polypeptide, which combine to form the functional protein, insulin.*

4. *Many polypeptide chains are not fully functional in their original states (directly from translation). Other structural groups, such as lipids, phosphate groups, or carbohydrate groups need to be added, often in multi step processes in the endoplasmic reticulum or Golgi. Changes to the polypeptide determine the protein's structure and function, or determine where the protein is transported to.*

5. *(see next column)*

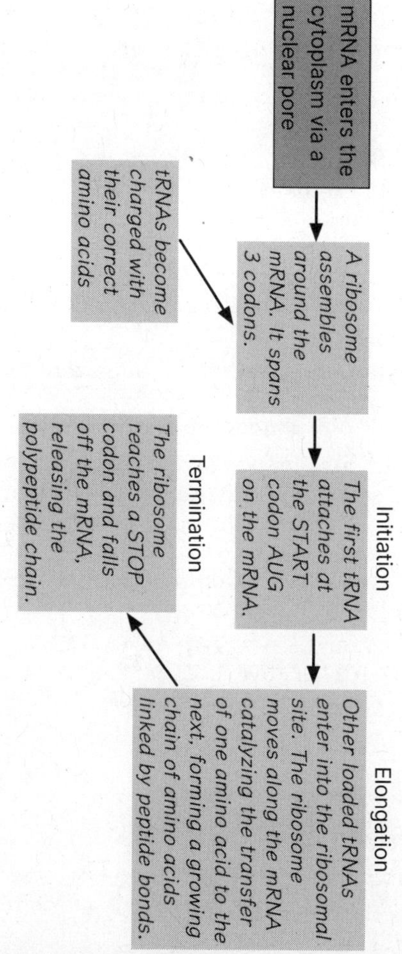

6 a. *tRNA need to be charged with their correct amino acids.*

b. *The ribosome (small subunit) needs to form around the mRNA.*

7. *Polypeptides are chains of amino acids bonded with peptide bonds (which are formed directly after translation. The final functional proteins have been modified, and bend/bonded into 3D structures with groups added.*

8. *The building blocks of the proteins, amino acids, are in constant demand to form new proteins that are required by the cells. Once the amino acids are released, they can be used over and over again to form different proteins.*

272. Gene Mutations (Page 432)

1 a. *A (reading) frame shift occurs when the sequence of bases is offset by one position (either by an insertion or deletion mutation), also called an INDELS mutation.*

b. *The frameshift alters the grouping of triplets for the entire sequence, which can severely alter the amino acid sequence in the protein.*

2 a. *Insertion or deletion (frameshift).*

b. *Rather than one single base change - which may or may not change an amino acid due to degeneracy, the frameshift (INDELS), has potential to change many amino acids in the chain - all of the codons after the point mutation are changed. This most likely results in a protein that is unable to bond correctly or be functional.*

ISBN: 978-1-99-101423-8

273. Causes and Consequences of Mutations (Page 433)

1. *A mutation is a change to the base sequence of an organism's DNA.*

2. *Mutations that benefit the individual by improving fitness will be retained (selection will favour them remaining in the gene pool). Mutations that are detrimental to fitness will reduce survival and reproductive success and so become less common in subsequent generations. Eventually, they will be eliminated.*

3 a. *A silent mutation is a point mutation that has no phenotypic effect,*
i.e. it does not change the protein encoded.

b. *Silent mutations can be carried without being subject to selection pressure. However, they do involve a change in the code and the base changes may be subject to different selection pressures and prove advantageous in a future environment.*

4 a. *Mutated DNA: CTG ATA AAA CTG GGG CTG TCA CCT AAT*

mRNA: GAC UAU UUU GAC CCC GAC AGU GGA UUA

Amino acids: Asp Tyr Phe Asp Pro Asp Ser Gly Leu

b. *Tyr - tyrosine*

c. *The mutation is silent. There is no change in the amino acid encoded and no change in the amino acid sequence.*

5. *Mutagens disrupt the DNA sequence in cells. The DNA is cleaved or otherwise damaged by the mutagens. Repair mechanisms in the cell may not be able to repair every disruption, leading to permanent changes in the DNA.*

6 a. *Mutation breeding is used to produce mutations that may lead to new and useful varieties of plants. It is also used to study the role of different genes.*

b. *The mutation is in the pollen and so is gametic (each pollen grain is a single cell containing two male gametes). The mutation will be present in the plants produced from this pollen.*

7. *Somatic mutations occur in the body (non-gametic or somatic) cells. They may affect an individual within its lifetime but are not inherited. Gametic mutations are mutations to the gametes (in testes or ovaries) and are inherited.*

8. *Somatic mutations will not be passed on (they are not heritable). Gametic mutations will be inherited. They can therefore become part of the genetic variation in the gene pool (upon which natural selection can act).*

9. *They can be used to develop better understanding of the effect of mutations to certain genes. In plants, chimeras can be used to produce new mutant plants by using cuttings (cloning cells in the region of the mutant). Mutants can then be studied, helping us to understand the role of the original gene.*

10. *A large gene pool has more alleles. If a particular environmental pressure arises or a disease enters the population, individuals possessing favourable alleles will survive and pass those to future generations, ensuring survival of the species.*

11. *Albino animals are more visible to predators and often do not survive in the wild long enough to breed as they are easily more preyed upon and killed. The mutation for the albino animal can only be passed onto the next generation if the animal survives long enough to have and rear offspring.*

12. *A bottleneck can vastly reduce the gene pool and number of alleles in a species. This gives the organism reduced resilience to possible disease or environmental pressures. Future generations have a reduced variety of alleles available to the population, e.g. cheetahs.*

13. *If information about a genetic disorder is known, such as the gene BRCA 2 which increases chances of developing breast cancer, then individuals may be able to take preventative measures. It can be concerning that commercial companies may hold potentially sensitive medical information that they may sell for a profit.*

274. Genetic Technology (Page 437)

1. *Developing mouse models for disease research in humans, Covid-19 testing, cure life threatening diseases, modify embryo cells with mutations, develop crops that can produce useful proteins or substances.*

2. *For example: Mouse models; human diseases can be studied without using humans - which can pose risk - especially if the disease is rare and experimentation may pose risk. The mouse models can be developed to replicate exact conditions that are faced by humans with the condition - in large numbers.*

3. *For example: Gene knockout has been used to disable genes in a leukaemia patient. By disabling particular genes (TCR-alpha and CD52), the patients were less likely to reject grafts introduced into the body - and make treatment more successful.*

4 a. *Student's answer*

b. *Students' responses will vary, but possible points include:*

PRO: Correcting mutations responsible for disease, switching faulty genes off so diseases or conditions do not develop; relatively low cost, relatively easy to carry out.

CON: Potential for off-target editing errors (other genes are adversely affected), Potential to create designer babies with specific traits, raises ethical questions about at what level of severity should a genetic problem be fixed? Not available to everyone due to cost. People may live longer and put pressure on Earth's resources.

275. Hypotheses for Conserved Genetic Sequences (Page 439)

1 a. *Highly conserved genetic sequences show very little change over time, with very few nucleotide changes, and even between distantly related species.*

b. *Some proteins synthesized from the genetic*

ISBN: 978-1-99-101423-8

sequences are essential to highly complex reactions; for example, those involved in cellular respiration. As any slight change can be fatal, it is essential that these remain unchanged.

2. Haemoglobin is on iron-containing protein found within red blood cells and is essential for the transport of oxygen around the body. It contains four subunits. Because it is an essential protein, the first hypothesis explains why it is so highly conserved between species and over time. Analysis of the amino acid sequence of haemoglobin between species shows it is a highly conserved protein. For example, humans and chimpanzees share the same amino acid haemoglobin sequence and there is only one amino acid difference between humans and gorillas. The differences are small, even between humans and other animals (27 between human-mouse and 67 between human-frog).

Any changes to the amino acid chain will affect the way the protein folds up (its three dimensional shape) and affect its functionality. Mutations resulting in changes to the amino acid sequence can cause serious and fatal health conditions, e.g. sickle cell disease. Therefore, because of the critical role of haemoglobin in oxygen transport, it is highly conserved.

276. Did You Get It? (Page 440)

1 a & b.

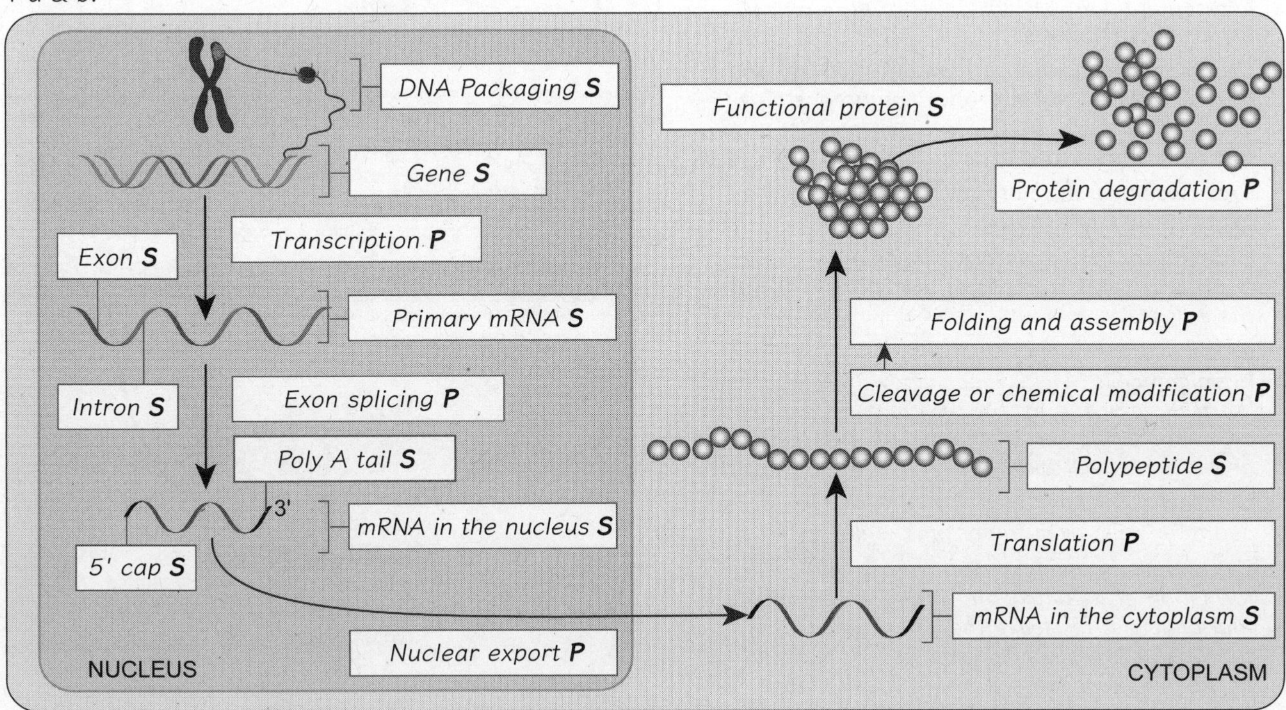

2. One new strand of DNA is synthesized in a 5' to 3' direction, from a template original strand - resulting in semi-conservative replication. Nucleotides are added through complementary base-pairing. The base pairing rule ensures that nucleotide A is always paired with nucleotide T and nucleotide C is always paired with nucleotide G. Incorrect replication can be prevented if bases do not bond (and proofreading).

3. The deoxyribose sugar that makes the backbone of the DNA molecule is asymmetrical. This gives the strand directionality. RNA polymerase can only synthesize the RNA (transcription) from 5' to 3' end. mRNA moves through the ribosome from 5' to 3' and is read in that direction.

4. If one SNP mutation occurs to change a nucleotide in a codon, it may not necessarily mean that the amino acid that is coded for will be different (most amino acids are linked to more than one similar codon) - so no protein change.

5. If the mutation leads to increased survival rate - and/or increased number of offspring carrying the mutated gene compared to those not carrying the mutation - this would cause the frequency of the mutated gene/allele to increase in the population's gene pool. Also, if those with the mutation travelled and established families away from the original village.

Theme D: Continuity and Change

Chapter 14: Cells

277. Cell Division (Page 443)

1 a. Mitosis occurs in body cells (somatic cells) in animals.

b. Mitosis is responsible for growth of an organism, repair and replacement of damaged cells, and for asexual reproduction in some eukaryotes.

2 a. Meiosis occurs in sex organs (testes and ovaries) in animals.

b. It produces sex cells (gametes) for the purposes of sexual reproduction.

3. Gametes are haploid (N) because meiosis halves the chromosome number of a somatic cell.

ISBN: 978-1-99-101423-8

Fusion of gametes in fertilization restores the diploid number for the organism (2N).

278. Mitosis and Cytokinesis (Page 444)

1. *The DNA must replicate.*

2 a. *The spindle fibres attach to the chromosomes to separate them and move them to opposite ends of the cell. They also cause cell elongation prior to telophase.*

b. *The spindle fibres originate from the centrioles of the centrosome.*

3. *Interphase*

3. *(Note: numbering sequence error)*

Dynamic movements (assembly and disassembly) of microtubular proteins are important in mitosis and cytokinesis (spindle fibre activity and regulating cytokinesis). These are energy expensive processes.

4 a. *Chromosomes condense (coil and fold up) and become visible. Nuclear envelope breaks down. The replicated centrosomes (each one containing two centrioles) move to opposite poles of the cell.*

b. *Chromosomes attach to the spindle fibres and align at the cell equator.*

c. *Chromatids from each chromosome are pulled apart and move in opposite directions, towards the poles.*

d. *Chromosomes begin to unwind. Two nuclei form. In animal cells, a cleavage furrow forms across the midline of the parent cell. In plant cells, a cell plate forms across the midline where the new cell wall will form.*

5 a. *Cytokinesis divides the cell to create two daughter cells from the original cell.*

b. *Cytokinesis in animal cells involves the formation of a contractile ring of microtubules that constrict to cleave the cell. In plant cell cytokinesis, vesicles deliver cell wall material to the middle of the cell where a cell plate (a precursor to the new cell wall) forms. The vesicles coalesce to form the plasma membranes of the new cell surfaces.*

6 a. *Anaphase*

b. *Prophase*

c. *Metaphase*

d. *Telophase*

7. *This ensures that one oocyte (egg cell) has enough material and energy to survive the journey through the fallopian tube to the uterus and to carry out the initial process of cell division after fertilization. If cell division was even, none of the cells would have enough material and energy to be able to survive this.*

8. *A cell's initial mitochondria come from the parent cell, as they are not manufactured by the cell. Subsequent mitochondria come from the growth and division of these initial mitochondria.*

279. Modelling Mitosis (Page 447)

Investigation 14.1 Modelling mitosis

3 a. *Chromosomes*

b. *Nuclear membrane*

c. *Plasma membrane*

4.

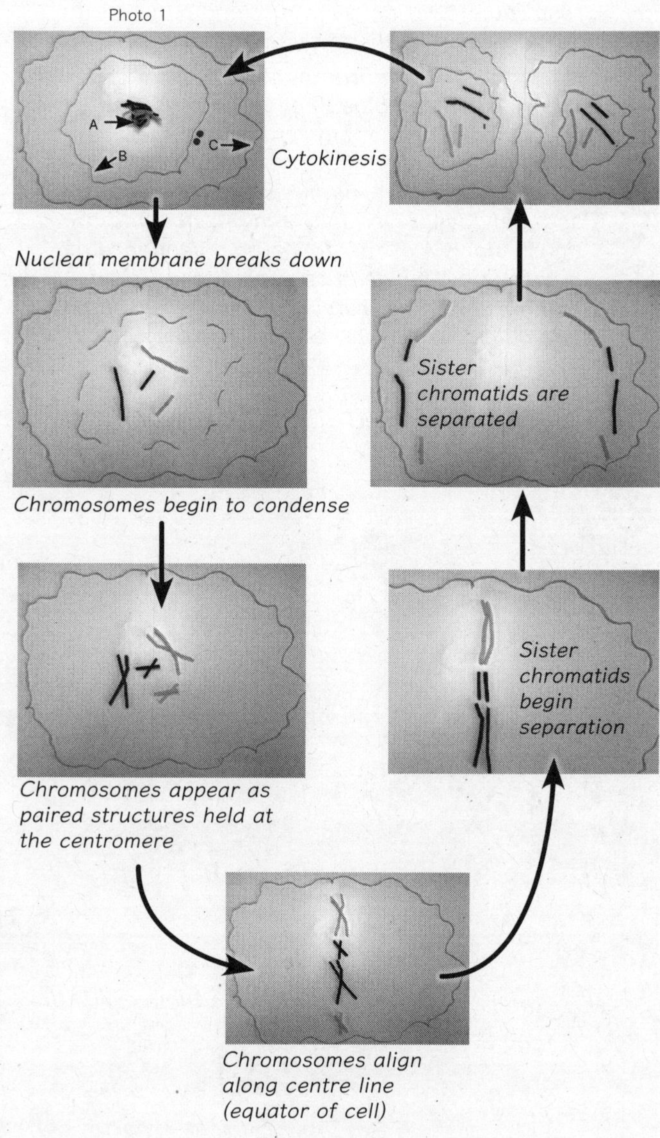

280. Meiosis and Variation (Page 448)

1. *In the first division of meiosis, homologous pairs of chromosomes pair to form bivalents. Crossing over may occur. Homologous pairs separate. In the second division, chromatids separate (are pulled apart), but the number of chromosomes stays the same. This is more or less a 'mitotic' division.*

2. *Independent assortment refers to the random distribution of maternal and paternal homologues to the gametes. This results in 2n possible combinations of maternal and paternal chromosomes in gametes, where n is the haploid number.*

3 a. *Crossing over is the mutual exchange of pieces of chromosome (alleles) between non-sister chromatids of homologous chromosome pairs.*

b. *Crossing over increases variation by creating new combinations of alleles on the chromosomes involved in the crossing over. The more crossing over incidents, the greater the genetic variation.*

ISBN: 978-1-99-101423-8

281. Nondisjunction in Meiosis (Page 450)

1. *Non-disjunction results in abnormal chromosome numbers in some gametes. As a result, certain phenotypic traits are exhibited (e.g. characteristic physical features, varying degrees of intellectual impairment). Certain metabolic processes may also be affected.*
2. *Non-disjunction in the parental cell during meiosis I results in both the daughter cells being faulty which will be transferred to the daughter cells produced in meiosis II. Nondisjunction in a parental cell during meiosis II results in only half the total number of daughter cells being faulty.*
3. *The maternal age effect is an increased risk of chromosome abnormalities with advancing maternal age.*

   The risk of aneuploidies increases with a woman's age.
4. *Half of the gametes would have no chromosome at all. The other half would have double the number of gametes. If one of the gametes join with a normal gamete, the offspring would have triple the normal number of chromosomes (polyploidy). In humans, this is not viable.*

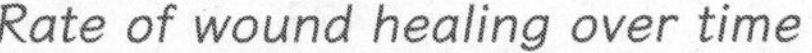

282. Growth and Repair (Page 451)

1 a.

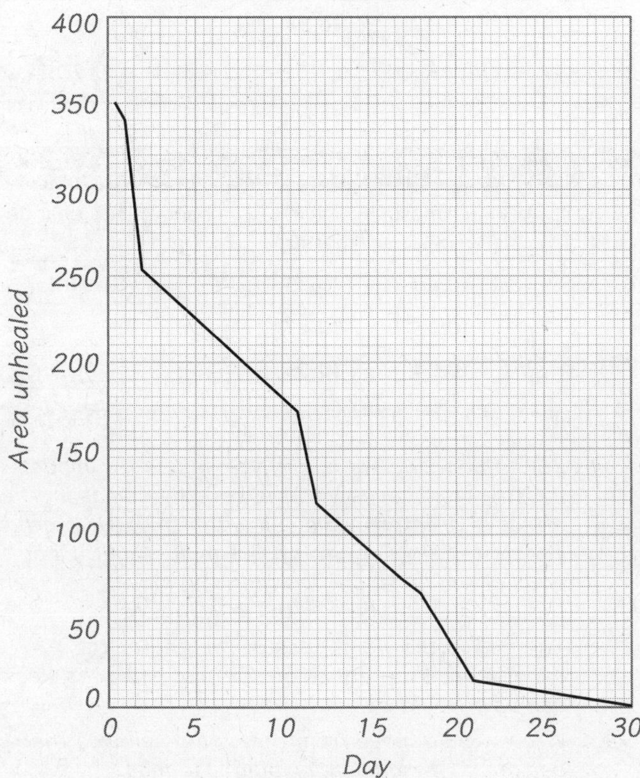

b. *The wound size drops in 3 major steps, with slower rate of healing between. So the rate cell division increases in bursts, rather than at the same rate.*

2. *The daughter cells produced by mitosis are identical, so they can replace cells or repair damage to the tissue that produced them (tissue of origin).*

283. The Eukaryotic Cell Cycle (Page 452)

1. *Interphase*
2. *S phase is the stage at which DNA/ chromosomes is replicated. M phase is the stage at which the replicated chromosomes are separated.*
3. *G1 is the stage at which proteins and material needed for DNA replication are made. G2 is the stage in which the cell prepares to divide (cytokinesis).*
4. *Cyctokinesis is the division of the cell.*

284. Regulating the Cell Cycle (Page 453)

1. *G1: Cyclin D,*
S phase: Cyclin E,
G2 phase: Cyclin A,
mitosis: Cyclin B
2. *Checkpoints ensure the cell is has the material and energy to move onto and complete the next stage, especially cell division and DNA replication.*
3. *The cell cycle may be shortened in response to damage, such as a cut or wound. Cells divide rapidly to heal the damage, then return to the normal rate of cell division.*

285. Disruption to the Cell Cycle (Page 454)

1 a. *Proto-oncogenes control the start of cell division. Tumour suppressor genes switch off cell division.*

The action of these two genes therefore regulate the timing of cell division.

b. *Normal controls over the cell cycle can be lost if either the proto-oncogenes or the tumour suppressor genes acquire mutations that prevent their correct function.*

c. *Mutations to the proto-oncogenes(and the consequent formation of oncogenes) results in uncontrolled cell division. Mutations to the tumour-suppressor genes results in a failure to regulate the cell repair processes and a failure of the cell to stop dividing when damaged.*

286. Determining the Rate of Growth (Page 455)

1 a. *58*

b. *7*

c. *0.12 (12%)*

2. *A higher than expected mitotic index would indicate cells dividing more than normal, which could indicate cancerous cells.*

3 a. *As effluent concentration increases, the rate of mitosis in onion cells decreases.*

b. *The root tip would not grow correctly, slow growth would occur.*

287. Gene Expression and Regulation (Page 456)

1. *Transcription factors control retinal cell fate. Different transcription factors cause the expression of different genes at different times and thus cause cell differentiation.*

ISBN: 978-1-99-101423-8

2. *This stops proteins being made when they are not needed. This tops resources being needlessly used, or overproduction that would cause problems in the cell.*

288. Epigenetics (Page 457)

1 a. *Histone modification may involve methylation or acetylation of histone tails. Acetylation causes looser packing of the DNA, whereas methylation causes tighter packing. DNA methylation causes chromatin to bind more tightly together.*

b. *Factors that cause the DNA to be more loosely packed make the DNA more accessible for transcription. Factors that cause the DNA to pack more closely together make the DNA unavailable for transcription.*

2. *Examples of the effect of environment on genes include the effect of air pollution and stress. Air pollution triggers epigenetic methylation in the Foxp3 gene which is linked with asthma. Studies using mice show these epigenetic changes appear to be heritable. Heightened stress has been shown to be linked to epigenetic changes. Children with mothers with PTSD tend to have lower than expected cortisol levels (stress related hormones).*

3. *The zygote is preprogrammed to cope with the environment that influenced the expression of genes in the parent, e.g. preprogrammed to survive famine conditions if the parent was subjected to a famine.*

4. *The mother must have donated the imprinted (silenced) gene because if the father's gene was working correctly there would be no syndrome as there would still be a working copy of the gene. If we know the father's gene was mutated and the syndrome appears, then the mother's corresponding gene must be silenced.*

5. *Because twins have essentially the same genes and DNA, studying twins who have lived apart gives an insight into how the environment affects genes and their expression.*

6. *The effect of environment on a person increases over a person lifetime. When young, a person's genes have more influence on their development. As a person is exposed to more things, the environment influences more genes and so a person's development is more influenced by the environment. E.g. a twin whose identical twin dies early has a high chance of also dying early, but if the twin dies at an older age the identical twin is less likely to also die.*

289. Gene Regulation (Page 460)

1 a. *Site of RNA polymerase binding to start transcription.*

b. *A non protein-coding sequence of DNA that is the binding site for the repressor molecule.*

c. *Genes responsible for producing enzymes that control the metabolic pathway.*

2. *The repressor molecule binds to the operator and switches off the gene.*

3 a. *Not attached*

b. *The effector activates the repressor which then binds to the operator.*

4. *When tryptophan is low, the operon is on. When tryptophan is high, the tryptophan combines with the repressor which can then bind to the operator. This blocks tryptophan production. When tryptophan is low, the repressor is released from the operator so more tryptophan can be produced.*

5. *Low oxygen signals to the body to produce more oxygen carrying blood cells so that the body's cells can obtain the oxygen they need. (This is why high altitude training increases a person's stamina at low altitude). When oxygen is plentiful, the body decreases blood cell production to replacement levels.*

6. *Environmental controls allow genes to quickly respond to environmental changes and switch on and off genes as they are needed, rather than being constantly on or off.*

290. Solutions (Page 462)

1 a.

1 molL^{-1}				
Cell	Final mass (g)	Initial mass(g)	change (g)	% change
1	**11.22**	**10.39**		
2	**11.23**	**10.33**		
3	**12.03**	**10.98**		
Mean	*11.49*	*10.57*	*0.92*	*8.70 %*

0.1 molL^{-1}				
Cell	Final mass (g)	Initial mass(g)	change (g)	% change
1	**10.44**	**10.35**		
2	**10.56**	**10.47**		
3	**10.64**	**10.55**		
Mean	*10.55*	*10.46*	*0.09*	*0.86%*

b. *Solute concentration is higher inside the model cells. Water moves into the cells by osmosis. The effect is greater in the cell with 1 molL-1 internal concentration than in the 0.1 molL-1 cell. Once equilibrium is reached, water still moves between the cell and the beaker but at equal rates.*

c. *The cell would weigh less at the end of the experiment than either of the first two cells (less osmosis would occur).*

d. *Water would move from the model cell into the beaker. The model cell would lose mass.*

e. *Water would move from the beaker into the cell, at around the some rate as the first model cell in the initial experiment (the difference is 1 molL^{-1} in both examples).*

2 a. *The solution is hypotonic to the cell. Water moves into the cell and it bursts.*

b. *The solution is hypertonic the cell. The cell loses water and becomes shrunken (crenulated).*

ISBN: 978-1-99-101423-8

c. The solutions inside and outside the cell are isotonic. The cell shape remains unchanged.

3 a & b.

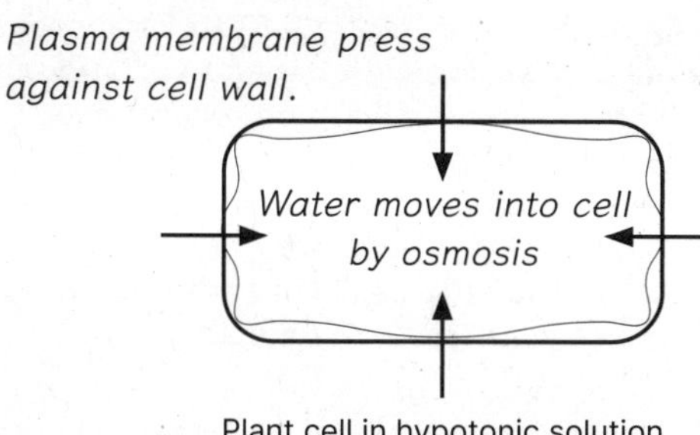

Plant cell in hypotonic solution

Plasma membrane pulls away from the cell wall.

Water moves out of the cell by osmosis

Plant cell in hypertonic solution

4. Isotonic

291. Estimating Osmolarity of Cells (Page 464)

1 & 2.

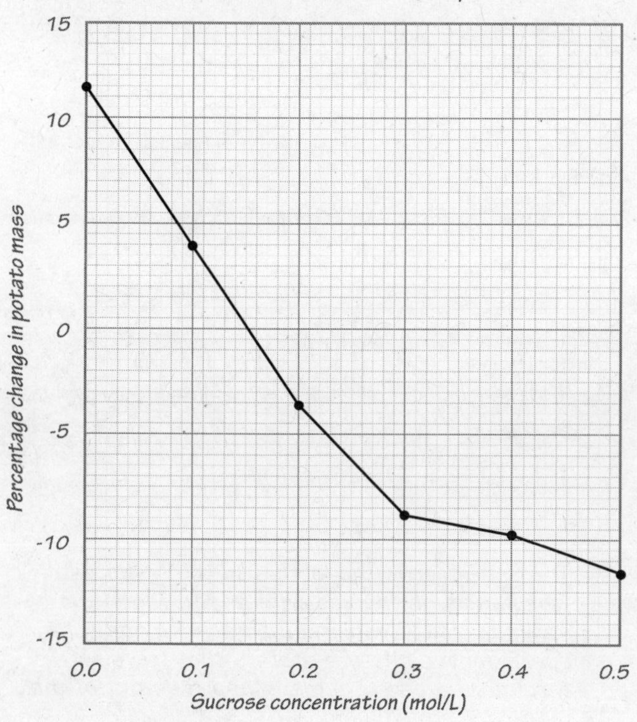

3. ~0.15 $molL^{-1}$

4. Hypotonic: 0.00 $molL^{-1}$ 0.1$molL^{-1}$
Hypertonic: 0.2 $molL^{-1}$, 0.3 $molL^{-1}$, 0.4 $molL^{-1}$, 0.5 $molL^{-1}$

5 a. The net movement of water is from the cell into the solution.

b. There is no net movement of water.

c. The net movement of water is from the solution into the cell.

292. Water Relations in Plants (Page 466)

1. Zero

2 a. ψ for side A: -100
ψ for side B: -200
Direction: A → B

b. ψ for side A: -400
ψ for side B: -500
Direction: A → B

c. ψ for side A: -400
ψ for side B: -200
Direction: A ← B

3. Dissolved solutes lower the water potential (make it more negative).

4. The plasma membrane pushes up against the cell wall, which is rigid, and stops the cell from bursting.

5 a. Plasmolysis is the pulling away of the plasma membrane from the cell wall, caused by a lack of water in the cell. Turgor is the pressing of the cell membrane against the cell wall as a result of water entering the cell.

b. The plant has wilted due to a lack of water. The cells in the plant have plasmolysed and lost their turgor, causing the plant to collapse.

6 a. Pressure potential generated within plant cells provides the turgor to support unlignified plant tissues.

b. Without cell turgor, soft plant tissues (soft stems and flower parts for example) would lose support and wilt.

293. Solute Potential and Cells (Page 468)

1 a. 1

b. 2

c. 3

2 a. -1 x 1 x 8.314 x 295 = -2452.6 kPa

b. -1 x 0.5 x 8.314 x 295 = -1226.3 kPa

c. -1 x 0.25 x 8.314 x 295 = -613.2 kPa

3.

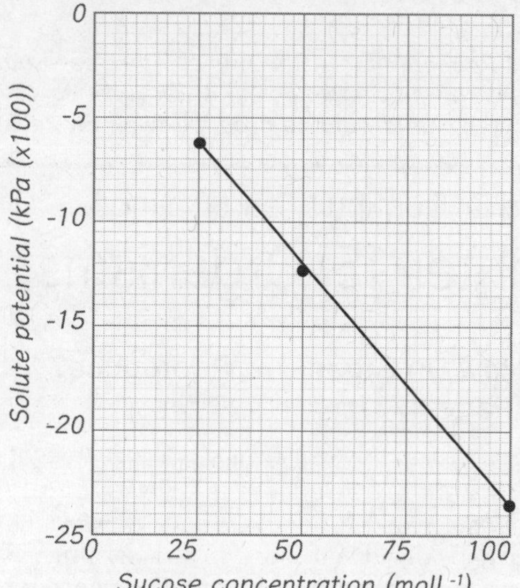

4. -1839 kPa

ISBN: 978-1-99-101423-8

294. Did You Get It? (Page 469)

1. *Meiosis*
2. *Meiosis produces gametes (in animals) for sexual reproduction (in plants, meiosis produces haploid spores).*
3. *Crossing over, in which genetic material is exchanged between non-sister chromatids of homologous chromosomes, and independent assortment, in which homologous chromosomes align randomly at the equator of the cell before being separated.*

4 a. *Mitosis*

b. *Growth and repair, asexual reproduction*

5. *Any animal cell can undergo mitosis (for tissue repair), in plants mitosis occurs in the meristems.*

6 a. *2*

b. *10*

c. *Yes*

7. *The salt produces a high salt concentration solution on the outside of the meat. Water moves from the interior of the meat in the salt, removing water from the meat (and any bacteria). The meat becomes so dry that no microbes can live on the meat.*
8. *From cell A to cell B*

9 a.

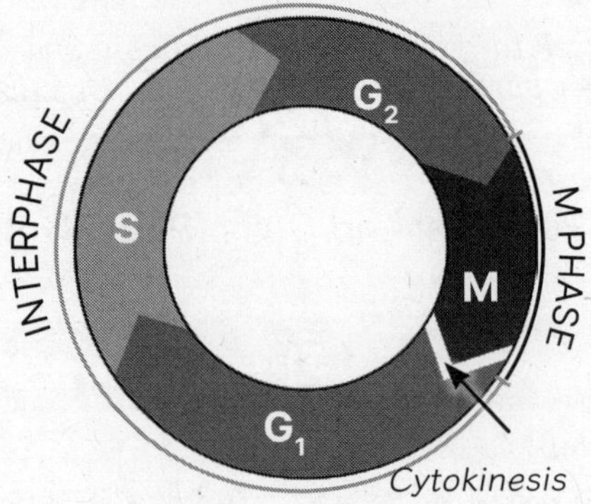

b. *G_1: Cell increases in size and makes the mRNA and proteins needed for DNA replication.*

G_2: Rapid cell growth and protein synthesis. Cell prepares for mitosis.

M Phase: Mitosis (the nucleus divides separating the chromosomal copies) and cytokinesis (the cytoplasm divides to produce two new cells).

Theme D: Continuity and Change

Chapter 15: Organisms

295. How Do Organisms Reproduce? (Page 473)

1. *The production of new life in order to continue the genetic lineage and continuity of the species.*
2. *Offspring produced by asexual reproduction are genetically identical to the parent. Offspring produced by sexual reproduction carry genetic material from both parents but are genetically distinct from them.*
3. *In a stable environment, asexual reproduction can be a benefit because it is simpler, safer, and produces genetically identical offspring which therefore have the same adaptations/traits/ phenotypes as the parent. Sexual reproduction produces offspring with genetic variation. This can be a benefit in changing environments, as some of the offspring may have phenotypes that better suit the changes than the parents, allowing for better survivability. The increase in frequency of the better adapted phenotypes results in evolution over time.*
4. *Meiosis reduces a germ cell from 2n to a gamete of 'n'. It introduces variation to the gametes, with a different combination of alleles compared to the parents.*
5. *The ovum is large compared to the sperm and for every round of meiosis, only one ovum is produced per germ cell. The ovum (one released each menstrual cycle) remains in the woman and contains enough energy to sustain growth through early fertilization. Contrastingly, the small sperm are produced in very large quantities. Four sperm gametes are produced each round of meiosis, and are produced continuously from puberty. Gametes contain just enough stored energy to move through the female reproductive system to reach the egg.*

296. The Male Reproductive System (Page 475)

1. *A front view of the male reproductive system is provided.*

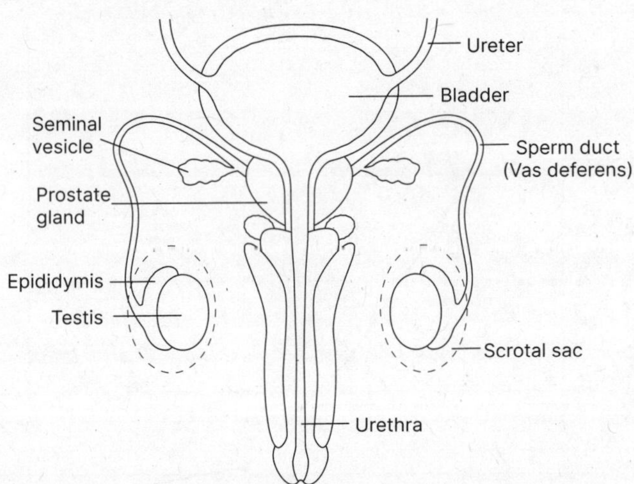

2. *The sperm leave the lobules in the testis and travel through the vas deferens to reach the seminal vesicle above the prostate gland. The sperm moves into the urethra that travels through the prostate gland and then out of the glans during ejaculation.*
3. *The tail enables the sperm to move independently through the woman's vagina, cervix, uterus, and then into the fallopian tubes to meet the egg. The energy for the movement is generated by the concentrated mitochondria wrapped around the head of the tail. The gametes are in the head, and the enzyme is used to penetrate the egg's outer membranes.*

ISBN: 978-1-99-101423-8

297. Spermatogenesis (Page 476)

1. *The testes house the seminiferous vessels and the epididymis. The seminiferous tubules in the testis are where sperm cells develop and the sertoli cells nourish the sperm. The epididymis is where the sperm cells undergo further differentiation and develop tails ready for release.*

2 a. *In the seminiferous tubules*

b. *Testosterone*

c. *The testis is located in the scrotum, which hangs outside of the core body and is therefore kept at a lower temperature than the body (skin conducts heat away).*

3. *Haploid. The sperm form through meiosis, therefore reducing chromosomes from 2n to n. This allows them to combine with n ovum to form 2n zygote during fertilization.*

298. The Female Reproductive System (Page 477)

1. *A front view of the female reproductive system is provided; clitoris omitted.*

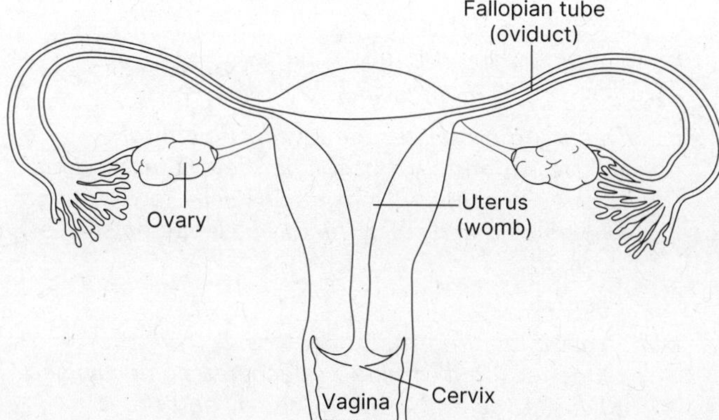

2. *The ovary is located at the top of the uterus and the ovum travels across to the fallopian tube. The ovum moves slowly down the fallopian tube and remains at body temperature the entire time. Testis are encased in the scrotum which hangs below the body to keep sperm cooler (20° C) to maintain viability and motility (movement).*

3. *Ovulation*

299. Oogenesis (Page 478)

1 a. *Female gametes/ovum start development in the embryo, which is suspended until puberty. At this stage, usually one ovum is fully matured and released each menstrual cycle. Sperm are produced continuously from puberty onward.*

b. *Women only have one opportunity to fertilize their egg each month (which stops during pregnancy) whereas males are able to fertilize as many ova as their sperm have access to - with no physical limits.*

2. *Fertilization of ovum.*

3. *Usually in the fallopian tube. However, sometimes fertilization occurs outside the reproductive system (ectopic) or once the ovum has reached the uterus.*

300. Fertilization in Humans (Page 479)

1. *The first sperm to reach the external membrane of the ovum fuses the membranes. This automatically prevents other sperm from gaining access to the egg. It is important that only one sperm (n) is able to fertilize one ovum (n) to form a 2n zygote - any other multiple of this would make the zygote non-viable. The nucleus of the sperm enters the cytoplasm of the egg and fuses with the ovum's nucleus to form a zygote.*

301. Fertilization and Early Growth (Page 480)

1 a. *Changes in the surface of the sperm cell (caused by the acid environment of the vagina) that make possible its adhesion to the oocyte.*

b. *The release of enzymes from the acrosome at the head of the sperm. These enzymes digest a pathway through the follicle cells and the zona pellucida.*

c. *Enables the sperm nucleus to enter the egg. The fusion causes a sudden depolarization of the membrane that forms a fast block to further sperm entry.*

d. *A permanent change in the egg surface that provides a slow block to sperm entry. Cortical granules are released into the perivitelline space, followed by the hardening of the vitelline layer.*

e. *The fusion of nuclei forms the diploid zygote and initiates the rapid cell division that follows fertilization.*

2. *It is necessary to prevent fertilization of the egg by more than one sperm because this would result in too many chromosomes in the zygote (making the zygote non-viable or unable to survive).*

3 a. *The oocyte is arrested in metaphase of meiosis II after it has undergone the first meiotic division.*

b. *Meiotic division is completed when the egg is fertilized.*

4 a. *Sperm contribution: 50%*
Egg contribution: 50%

b. *Sperm contribution: 0%*
Egg contribution: 100%

5. *Cleavage is the rapid early cell division of the fertilized egg to produce the ball of cells that will become the blastocyst. Cleavage increases the number of cells but not the size of the zygote.*

6 a. *Implantation establishes close contact between developing foetus and uterine lining, providing for the early nourishment of the embryo.*

b. *HCG prevents degeneration of the corpus luteum, so that it continues to secrete progesterone and maintain the pregnancy (placenta takes over this role at 12 weeks).*

7. *This is the period when most organ development occurs so the tissues are most prone to damage from drugs.*

ISBN: 978-1-99-101423-8

302. The Menstrual Cycle (Page 482)

1 a. *FSH*

b. *LH (surge)*

2. *LH and FSH levels stimulate oestrogen release, secreted by the Graafian follicle. Oestrogen peaks in levels around 12 days and triggers ovulation (ovum releases) which also results in a LH surge shortly after.*

3. *The growing follicle secretes oestradiol which stimulates the formation of the thickening uterus wall. This then suppresses LH and FSH - if the ovum is then fertilized, then this suppression maintains the thick uterus lining and prevents further ovulation. If ovum is not fertilized, the uterus wall is shed and ovulation occurs in the next cycle.*

303. Hormones and Puberty (Page 483)

1. *The release of GnRH from the hypothalamus at the start of puberty. This leads onto release of LH stimulating testis cells to secrete testosterone and FSH to stimulate ovaries to release oestrogen - both of which then lead onto development of secondary sex characteristics - and maturity of gametes.*

2. *To increase the width so that the baby (specifically head and shoulders) can be birthed.*

3. *Results in sex hormones being produced in sex organs = stimulate the development of secondary sex characteristics.*

4. *Primary sexual characteristics are the distinguishing characteristics that are either male or female (i.e. penis and testes, or vagina, uterus, and ovaries). Secondary sexual characteristics are male or female characteristics that develop after puberty under the influence of reproductive hormones.*

304. Treating Infertility with In Vitro Fertilization (Page 484)

1 a. *Halts the current menstrual cycle to 'reset' the process.*

b. *Stimulates the maturity of multiple ova - given in concentrations above the normal ovulation rate.*

c. *Levels of this hormone can be used to indicate when the ova are matured and ready to be harvested.*

d. *Induces synchronized ova release so that artificial extraction can occur.*

305. Detecting Pregnancy (Page 485)

1. *Gives an instant result and may circumvent a costly visit to a doctor until a pregnancy is confirmed. For some people, pregnancy detection in the privacy of their home is preferred.*

2 a. *Two*

b. *If hCG is present it will bind to mobile hCG-Ab. This complex will move by capillary action to a line of immobilized hCG-Ab and bind to them, producing a line. A second line will appear even if there is no hCG present and bind to hCG-Ab. Unbound hCG-Ab moves past the first line of immobilized hCG-Ab to a second line of anti-hCG-Ab, producing a second line. This line confirms the test is working.*

3. *Progesterone is also produced by the mother, so any detection of progesterone would not distinguish between the mother and the foetus.*

306. The Placenta (Page 486)

1. *A double layered, spongy, vascular tissue formed from foetal and maternal tissues in the wall of the uterus. The foetal portion of the placenta produces villi that project into the maternal endometrium which contains capillaries that connect the foetal arteries and vein. The blood vessels of the mother and foetus form a complex network for the exchange of nutrients, wastes.*

2 a. *~~Oxygenated and containing nutrients~~ / Deoxygenated and containing nitrogenous wastes*

b. *Oxygenated and containing nutrients / ~~Deoxygenated and containing nitrogenous wastes~~*

307. The Hormones of Pregnancy (Page 487)

1 a. *The corpus luteum acts as the source of progesterone to maintain the endometrium before the development of the placenta.*

b. *Progesterone and oestrogens.*

2 a. *Oestrogen (high levels) and oxytocin.*

b. *Declining progesterone levels. Placental deterioration. High oestrogen levels increasing uterine sensitivity to oxytocin. Peak in oxytocin. Physiological state of foetus (release of stress hormones etc.).*

3 a. *Positive*

b. *Prostaglandins, factors released from the placenta (as its function becomes compromised towards the end of pregnancy), and the physiological state of the baby itself.*

4. *Oestriol, oxytocin.*

5. *The birth of the baby.*

308. Hormone Replacement Therapy (Page 489)

1. *Because the rate of CHD was measurably lower in the women in the HRT experimental group than those women who did not take HRT. An assumption was made between the causal link of the two variables.*

2. *The control (not taking HRT) could be measured within the clinical environment - with every other variable also controlled (demographics, including health and financial ability to purchase HRT treatment). The randomized selection of groups reduced the chance of bias - i.e. inadvertently selecting more healthier women for groups.*

309. Sexual Reproduction in Flowering Plants (Page 490)

1. *Plants cycle between a diploid, sporophyte generation (produces haploid spores by meiosis) and a haploid gametophyte generation (produces the haploid gametes by mitosis).*

ISBN: 978-1-99-101423-8

2 a. Anther

b. Anther

3.

Structure	Spores	Gametes	Zygote
Produced by	Sporophyte	Gametophyte	Gamete fusion
Process	Meiosis	Mitosis	Fertilization

310. Insect-Pollinated Flower Structure (Page 491)

1. *A labelled diagram of an insect pollinated flower is below. Male structures are identified with (M) next to the label and female structures with (F). Functions of the structures are provided below the diagram.*

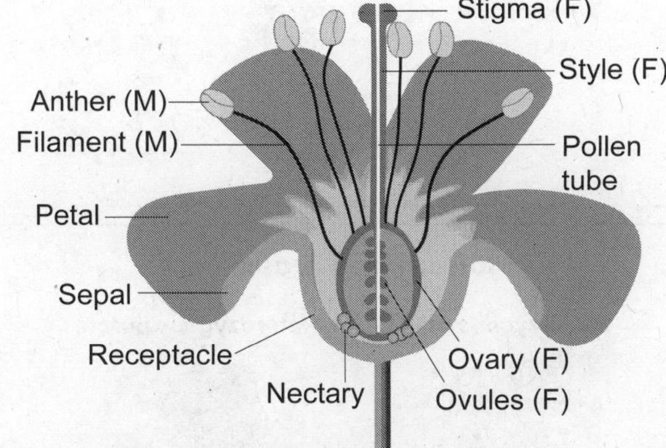

Function of labelled structures:

Stigma (F): *The receptive part of the carpel. Pollen grains will germinate here.*

Style (F): *The structure that supports the stigma.*

Ovary (F): *An enlarged base of the carpel. The ovules develop within the ovary.*

Ovules (F): *These become the seeds once fertilization has occurred.*

Anther (M): *Top portion of the stamen, the male organ of reproduction where pollen is produced.*

Filament (M): *The slender stalk of the stamen that supports the anther.*

Petal: *Surround the reproductive structures of the flower, petals are often brightly coloured to attract insects.*

Sepal: *Forms part of the flower, acts to protect the flower. Collectively the sepals are called the calyx.*

Nectary: *A depository for nectar. Nectar attracts pollinating insects to the flower.*

Receptacle: *The swollen base of the flower.*

2 a. *Flowering plants need to attract pollinators to collect pollen from one flower and move it to another, preferably on a different plant, and of the same species.*

b. *Flowering plants attract pollinators by using flowers (which may have scents and colours detectable by insects) as attractants. Most offer nectar as a reward for a pollinator. Many produce large amounts of pollen which can also act as a reward for pollinators.*

311. Cross-pollination and Fertilization Mechanisms (Page 492)

1. *Limiting self pollination prevents the negative effects of inbreeding and increases variability and therefore adaptability in a changing environment. Self-pollination results in the offspring all being identical.*

2. *Both forms of pollination allow the pollen to move away from one plant and potentially towards another. Animal pollination is likely to directly target the flower of the same species plant, but comes at a cost to plants to develop flowers and nectar. Wind pollination is more random, and pollen will land mainly away from the flowers of plants of the same species.*

3. *Pollination refers specifically to the transfer of pollen from the male anthers to the female stigma. Fertilization in plants refers to formation of the embryo by the fusion of a sperm nucleus with the egg.*

4. *The diversity of the plant gene pool is reduced - genetically identical plants have less capacity for survival if the conditions change - reducing opportunity for evolution. It can also increase the chances of mutations - if they are rare recessive mutations.*

5. *Other methods of self-incompatibility in plants students may study include:*
   - *S-glycoprotein mechanism*
   - *2-locus gametophytic self-incompatibility*
   - *Heteromorphic self-incompatibility*
   - *Cryptic self-incompatibility (CSI)*
   - *Late-acting self-incompatibility (LSI)*

312. Seed Dispersal (Page 494)

1 a. *Wind*

b. *Very light seeds with feathery heads to capture the air currents.*

2 a. *Germinate where they fall.*

b. *High energy, readily accessible food for forest animals. Acorns that escape predation have a better chance of vigorous growth. Some animals may bury the acorn thus enhancing chances of germination.*

3 a. *Water*

b. *The fruits have thick fibrous walls containing air pockets, enabling them to float. The seed inside remains unaffected by the salt water and germinates once it is washed ashore.*

4 a. *Wind*

b. *Large seeds attached to 'wings' which act like a propeller in the air currents, carrying the seed away from the parent plant.*

5 a. *Animal*

b. *The fleshy structures surrounding the seeds are*

ISBN: 978-1-99-101423-8

eaten by insects such as ants, and the seeds are then abandoned.

6 a. Animal

b. High energy fruits are favoured by birds and mammals. The seeds are dropped or pass through the digestive tract to be deposited elsewhere.

313. Seed Structure and Germination (Page 495)

1. Pollination is the dispersal of the gametes - typically just the male gamete in the form of pollen. Seed dispersal is the dispersal that occurs to the seed structure once which occurs after fertilization. Pollination occurs to the pollen during the gametophyte stage of the plant, while seed dispersal occurs during the sporophyte stage.

2. A seed uses a food store, produced by mitosis during seed development, because there is a period of growth (which requires energy) before leaves develop and, therefore before any photosynthesis occurs.

314. The Genetic Basis of Inheritance (Page 496)

1. Mendel had proposed his laws of inheritance but did not yet understand the physical mechanism that allowed them to occur - genes, as segments of coded information in DNA, and contained in chromosomes, supported his laws.

2. Almost all cells in the body are somatic cells and in humans contain 46 chromosomes (2n) in 23 homologous pairs. Haploid cells are gametes found in the reproductive organs - and formed during meiosis, have just 23 chromosomes (n) in humans. Both cells have all the common cellular components however, such as a membrane, nucleus, and mitochondria.

3. The gametes contain one half of the full set of chromosomes - when the two gametes (sperm and ovum) combine then they both contribute one chromosome from each homologous pair. Therefore, the zygote contains the full 46 chromosomes (2n) - the cell, as it continues to divide, now carries one allele from each parent, for each gene.

315. Genotype and Alleles (Page 497)

1 a. Genotype contains one dominant and one recessive allele.

b. Genotype contains 2 dominant alleles.

c. Genotype contains 2 recessive alleles.

2 a. Aa

b. AA

c. aa

3. Chromosomes that contain the same genes in the same position (loci), but the genes may exist as different versions (alleles) or be the same alleles (homozygous). Each chromosome of the homologous pair came from a gamete (mother and father) and combined during fertilization.

4. Because most genes can exist as alleles, different varieties of the same gene can have slight differences in the base combinations - this can possibly be expressed as different phenotypes - due to different polypeptide products being produced. Organisms with different phenotypes (from the alleles) may have differential survival rates, and lead onto greater fitness.

316. Genetic Crosses in Flowering Plants (Page 498)

1.

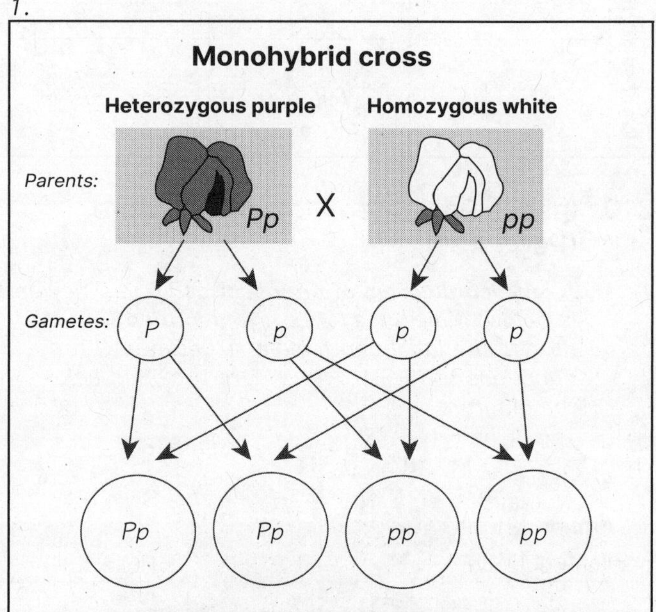

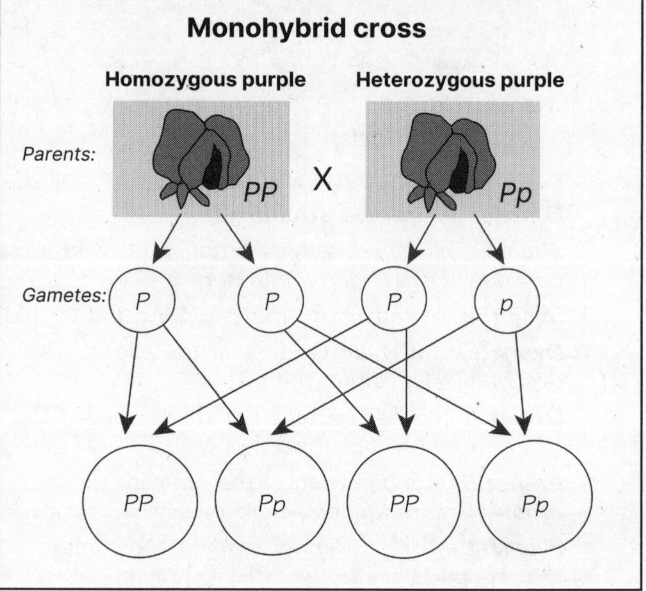

2. A dominant trait (produced by a dominant allele) overrides a recessive trait and only one copy of the allele needs to be present for the trait to be expressed. A recessive trait is only expressed when there are two copies of the recessive allele.

3. The wrinkled seed trait must have been masked by the smooth seed trait, i.e. it was recessive. The F_1 generation must have had both parents having heterozyogous genotypes: white flower phenotype has a pp genotype, which means that both parents needed to contribute a recessive allele each.

ISBN: 978-1-99-101423-8

4 a &b.

Dominance	Ratio
Dominant: **Round** Recessive: **Wrinkled**	**2.96 : 1**
Dominant: *Yellow* Recessive: *Green*	*3.01:1*
Dominant: *Green* Recessive: *Yellow*	*2.82:1*
Dominant: *Axial* Recessive: *Terminal*	*3.14:1*
Dominant: *Constricted* Recessive: *Inflated*	*2.95:1*
Dominant: *Tall* Recessive: *Dwarf*	*2.84:1*

5. *Mendelian crosses are only predictions - in a much larger sample group, the ratios are likely to approximate the 3:1 ratio closer.*

317. Phenotype and the Environment (Page 500)

1. *Students' own data.*

   Students can collect the raw data on a tally chart, and produce a bar graph to display the results.

   They should see that the recessive trait shows a lower frequency.

2. *Student answer: Could be hair blonded by sun exposure, muscle strength, freckles from the sun, a tanned skin colour etc.*

318. Variation and Phenotypic Plasticity (Page 501)

1. *A surplus of female fish allows for a greater reproductive rate as only one male in a small population is able to fertilize the females. So, rather than having an excess of males, being able to just allow a female to 'morph' into a male when required is the most efficient method of having the most individuals who can reproduce.*
2. *The helmet may be a disadvantage to the Daphnia when gathering food, reproducing, or moving around. If the helmet is not required, then being able to 'lose' the helmet when there are no predators around is an advantage.*
3. *In the dry season, the vegetation tends to be a brown colouration, so the duller butterfly is able to blend in and camouflage, especially when it is not actively out eating all the time. In the wet season when food is plentiful and it is out eating more constantly, the butterfly is able to take advantage and reproduce at a much faster rate. The more visible colouration may become more important to attracting mates and also warn against predators.*

319. Phenylketonuria (Page 502)

1. *In an autosomal recessive inheritance pattern, a person can be affected or unaffected, but also be a carrier and unaffected because the trait will be expressed only if there are two alleles present.*
2. *If a person is unaffected, when the trait is dominant then they will have no mutated allele, whereas with a recessive trait, they may be carriers. This mutation may show up in the next generation if each parent passes on a mutated allele. An affected person needs only one mutated allele of the trait to be dominant.*
3. *Rare mutated alleles are not common in a population but are passed down to related individuals. If two closely related individuals have offspring, then there is a much greater chance that they both have a mutation - and if the disease is autosomal recessive, the parents may be unaware they are carriers until a child is born.*
4. *The heelprick is a simple and easy test - the disease is not usually discovered without this test until symptoms develop. The disease is easy to control, with few symptoms if it is discovered at an early stage.*

320. Alleles in Populations (Page 503)

1. *Even though a fertilization only combines 2 alleles, if the options for a greater range of alleles is present in a gene pool, then the possible combinations of alleles in the offspring is greater. This will mean that there is greater diversity in the population which gives the population an advantage if environmental conditions change.*

321. Multiple Alleles and Codominance (Page 504)

1. *Both alleles in a heterozygote are equally and fully expressed and this produces a new phenotype. So, $R^w R^w$,*

   $R^w R^r$, and $R^r R^r$ genotypes each have their own distinct phenotype.

2 a.

I^A I^B I^A I^B

$I^A I^A$ $I^A I^B$ $I^A I^A$ $I^B I^B$

A AB A B

b.

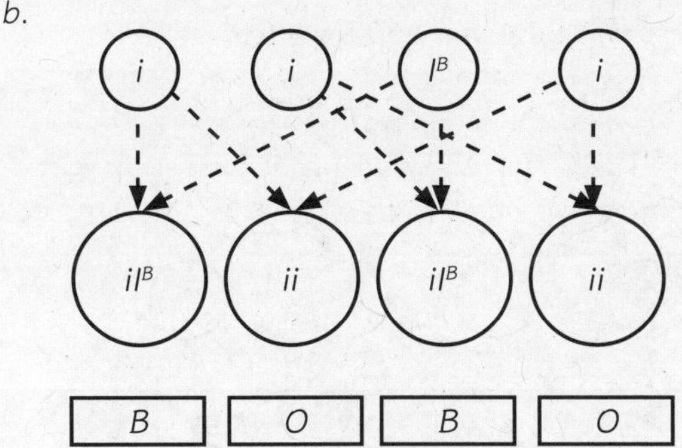

ISBN: 978-1-99-101423-8

c.

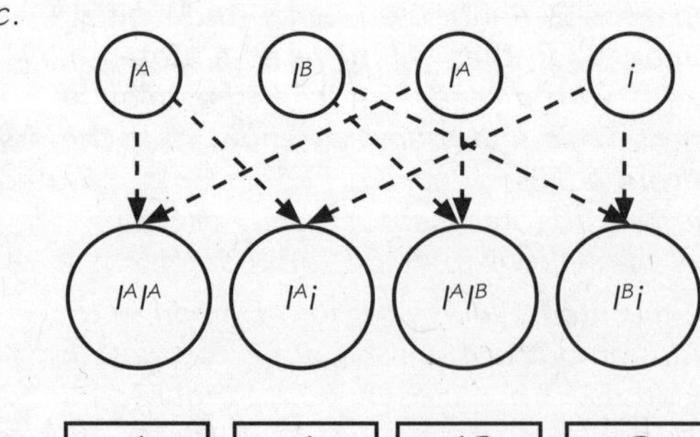

A	A	AB	B

d.

I^A I^A I^B i

I^AI^B I^Ai I^AI^B I^Ai

AB	A	AB	A

3 a.

R^rR^r R^rR^W

R^r R^r R^r R^W

R^rR^r R^WR^r R^rR^r R^WR^r

b. Red or ~~roan~~

4. CRCW X CRCW gives CRCR : CRCW , CRCW : CWCW

Ratio: 1 : 2 : 1 (1 red : 2 roan : 1 white)

322. Incomplete Dominance (Page 506)

1. In incomplete dominance, neither allele is fully dominant and both are partly expressed, so the phenotype is intermediate. In complete dominance, the dominant allele will override the recessive one and the dominant phenotype will be expressed.
2. Parents: white and pink.

   Offspring 50% pink, 50% white.

3 a.

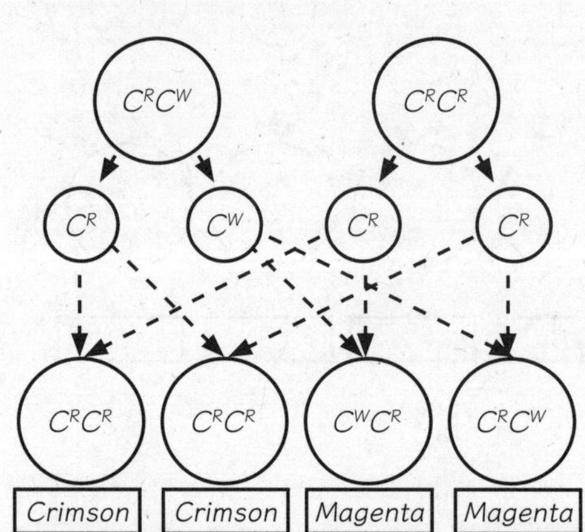

b. 50% crimson, 50% magenta (1:1)

323. Sex Determination and Linkage (Page 507)

1 a & b.

Male children	X^HY	Normal
	X^hY	Haemophiliac
Female children	X^HX^H	Normal
	X^HX^h	Carrier

2 a. Maternal GF must be XhY so the mother must be a carrier (see squares). 50% chance woman is carrier.

b. If a woman is normal then 0%

If a woman is a carrier then 50%

3 a.

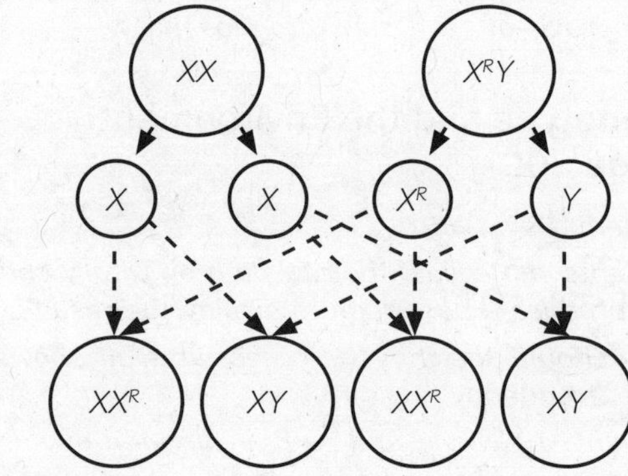

b. 50%

c. 100%

d. 0%

4 a.

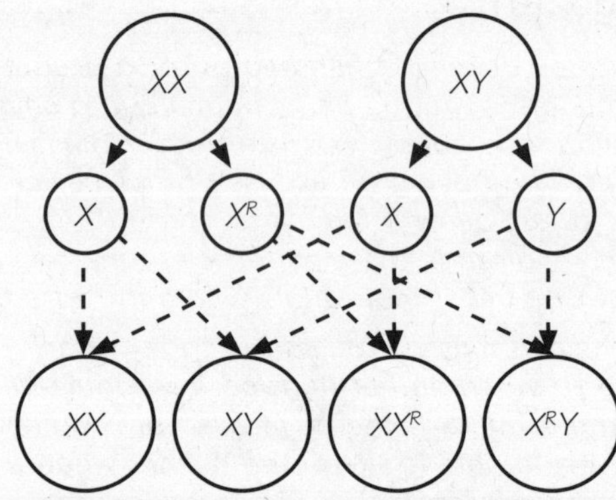

b. 50%

c. 50%

d. 50%

5. Because they have only one locus for the gene, they must express the trait. If the trait is recessive, females will express the phenotype only when homozygous recessive. Females can inherit a double dose of the recessive allele, but this is much less likely than in males because sex linked traits are relatively uncommon.

6 a. No females have the disorder

b. The male parent must be affected.

ISBN: 978-1-99-101423-8

324. Pedigree Analysis (Page 509)

1 a.

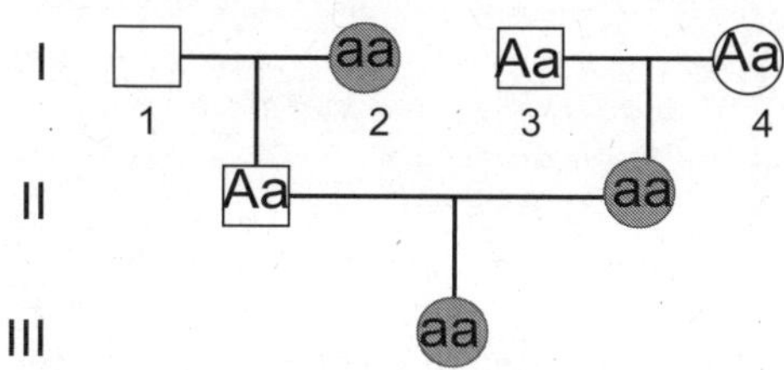

b. I-3, I-4, II-1

c. I-1

d. 0.25

e. Man (A) must be heterozygous because he has a blue eyed child. There is always a 50% chance of a child with his blue-eyed partner (B) being blue eyed, so probability of children being blue-eyed is 0.5. The probability of child being a boy is also 50% or 0.5. 0.5 x 0.5 = 0.25

2. Lactose intolerance must be autosomal recessive.
Evidence 1: It is recessive because it requires two copies of the faulty gene to occur (two lactose tolerant parents can produce a lactose intolerant child. For example III 3&4 and III 5&6 are both lactose tolerant, but both produce children who are lactose intolerant (IV 5 and IV 6).
Evidence 2: It is not sex linked because females and males show lactose intolerance.

3 a. 50% Li and 50% ll

b. He has to be heterozygous (a carrier) because if he was homozygous dominant none of his children would be lactose intolerant. Therefore, his genotype is Ll.

325. Polygenes (Page 511)

1 a. Discrete variables are either/or: this is seen in phenotypes as either having the trait or not. In continuous variables, the values can be anywhere along the scale between both extremes: this is seen as a trait that is somewhere between, or at one end.

b. Many genes allows for many combinations of alleles between the most 'dominant' dihybrid homozygous genotype and the most 'recessive' dihybrid homozygous genotype - the phenotypes in between are varying combinations of the two genes, with varying continuous traits as a result.

2. 1:6:15:20:15:6:1.
Black and light/pale have the lowest frequencies, while medium skin colours have the highest.

Investigation 15.1: Measuring continuous variation

1. Student's or class choice - the investigation is easier if one variable chosen for entire class

2. Raw data (e.g. weight in kg)

63.4	68.5	81.2	76.0	65.0	66.0
56.5	67.2	83.3	72.5	75.6	65.5
84.0	82.5	95.0	61.0	76.8	67.4
81.5	83.0	105.5	60.5	67.8	73.0
73.4	78.4	82.0	67.0	68.3	67.0
56.0	76.5	73.5	86.0	63.5	71.0
60.4	83.4	75.2	85.0	58.0	70.5
83.5	77.5	63.0	93.5	58.5	65.5
82.0	77.0	70.4	62.0	50.0	68.0
61.0	87.0	82.2	62.5	92.0	90.0
55.2	89.0	87.8	63.0	91.5	83.5
48.0	93.4	86.5	60.0	88.3	73.0
53.5	83.0	85.5	71.5	81.0	66.0
63.8	80.0	87.0	73.8	72.0	57.5
69.0	76.0	98.0	77.5	66.5	76.0
82.8	56.0	71.0	74.0	61.5	

3. Tally chart

Weight group (kg)	Tally	Total
45-49.9	I	1
50-54.9	II	2
55-59.9	IIII II	7
60-64.9	IIII IIII III	13
65-69.9	IIII IIII IIII	15
70-74.9	IIII IIII III	13
75-79.9	IIII IIII I	11
80-84.9	IIII IIII IIII I	16
85-89.9	IIII IIII	9
90-94.9	IIII	5
95-99.9	II	2
100-104.9		0
105-109.9	I	1

4. Number of entries: 95
Mean: 73.7*
Median: 73.4*
Mode: 80 - 84.9*
*calculations provided above are based on the data provided in the example set.

5.

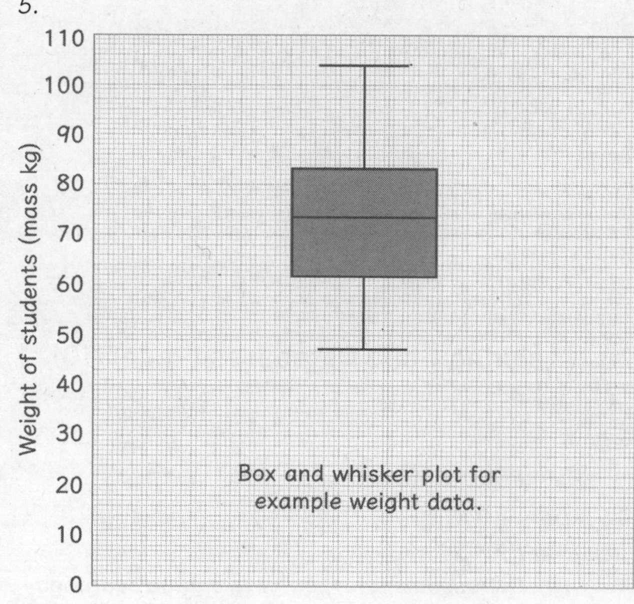

ISBN: 978-1-99-101423-8

3 a. *Students should describe the data as showing a normal distribution (with closer Q1 and Q3 boxes than the 'whisker' from Q3 to minimum and Q3 to maximum) because the data collected show continuous variation.*
If the data do not appear as a normal distribution, the students should accurately describe the shape of the box and whisker plot.

b. *The genetic basis for this distribution is polygenic inheritance.*
This is where several genes are involved in determining the phenotypic trait so there are many possible outcomes. The environment may also have an influence, especially if traits such as weight are chosen.

4. *Traits with continuous variation show a normal distribution when sampled and a graded variation in phenotype in the population. Such phenotypes are usually determined by a large number of genes and/or environmental influence, e.g. height, weight, hand span, foot size. Traits with discontinuous variation fall into one of a limited number of phenotypic variants and do not show a normal distribution curve when sampled, e.g. chin cleft.*

326. Mendel's Laws of Inheritance (Page 514)

1. *Law of particulate inheritance.*
2. *Law of independent assortment.*

327. Dihybrid Inheritance (Page 515)

1 a.

	BL	Bl	bL	bl
BL	*BBLL*	*BBLl*	*BbLL*	*BbLl*
Bl	*BBLl*	*BBll*	*BbLl*	*Bbll*
bL	*BbLL*	*BbLl*	*bbLL*	*Bbll*
bl	*BbLl*	*Bbll*	*bbLl*	*bbll*

b. *1 - BBLL*
2 - BbLL
2 - BBLl
4 - BbLl
1 - BBll
2 - Bbll
1 - bbLL
2 - bbLl
1 - bbll

328. Exploring Gene Databases (Page 516)

1 a. *99%*

b. *92%*

2. *1 (human) and 90 (wolf)*

3 a. *99%*

b. *726 (Human: A, dog: T)*

4. *Any DNA sequence typed here can be searched for in the database. Thus, any newly found DNA can be searched for a match. If no match occurs, the DNA may be from a new species or, if a closely related species has been sequenced earlier, it will show up in the search.*

5. *Student's own answer. Answer will depend on the sequence they generate and the BLAST results obtained.*

6. *The ability to search for sequences and determine the relatedness of organisms is very powerful when it comes to identifying new organisms and their evolutionary relationships, as well as supporting or refuting currently accepted relationships (which may have applications to species conservation). Bioinformatics is particularly useful when analysing genomic information from new pathogens as it can provide insights as to the origins of those pathogens and so (potentially) the most effective course of prevention or treatment.*

329. Inheritance of Linked Genes (Page 518)

1. *Linkage refers to the situation where genes are located on the same chromosome. As a result, the genes tend to be inherited together as a unit.*

2. *Gene linkage reduces the amount of variation because linked genes tend to be inherited together and fewer genetic combinations of their alleles are possible.*

330. Recombination and Dihybrid Inheritance (Page 519)

1. *Because linked genes tend to be inherited together, there is less chance that crossing over will occur to split the genes (alleles) and more of the offspring will be the same as parental types (fewer recombinants).*

331. Testing the Outcomes of Genetic Crosses (Page 520)

1 a. *"If both parents are heterozygous and there is no linkage then we would expect to see a 9:3:3:1 ratio of phenotypes in the offspring ".*

b. *"If both parents are heterozygous and the genes are linked, then we would expect the ratios of phenotypes in the offspring to deviate from the 9:3:3:1"*

2 a.

Category	O	E	O-E	$(O-E)^2$	$\frac{(O-E)^2}{E}$
Purple steam jagged leaf	*12*	*16.3*	*-4.3*	*18.5*	*1.1*
Purple steam smooth leaf	*9*	*5.4*	*3.6*	*13*	*2.4*
Green steam jagged leaf	*8*	*5.4*	*2.6*	*6.8*	*1.3*
Green steam smooth leaf	*0*	*1.8*	*-1.8*	*3.2*	*1.8*
	Σ29				*Σ6.6*

b. *6.6*

c. *(4-1=) 3*

d. *The critical value is 7.82.*

e. *Circle "Do not reject H_0"*

3 a. *H_0 and H_A as for question 1.*

ISBN: 978-1-99-101423-8

b. 2.03
c. (4-1=) 3
d. Circle "Do not reject H_0"
e. In both cases, we cannot reject H_0, but in the first case, the χ^2 value is much higher. In tomatoes, the genes for stem colour and leaf shape are on separate chromosomes, but given the relatively large χ^2 value, repeating the experiment with more plants, or replicates, would serve as a check.

332. Principles of Homeostasis (Page 522)

1. Homeostasis is the relatively constant internal state of an organism, even when the external environment is changing.

2 a. Detects a change in the environment and sends a message (electrical impulse) to the control centre.
b. Receives messages sent from the receptor, processes the sensory input and coordinates an appropriate response by sending a message to an effector.
c. Responds to the message from the control centre and brings about the appropriate response, e.g. muscle contraction or secretion from a gland.

333. Negative Feedback Loops (Page 523)

1.

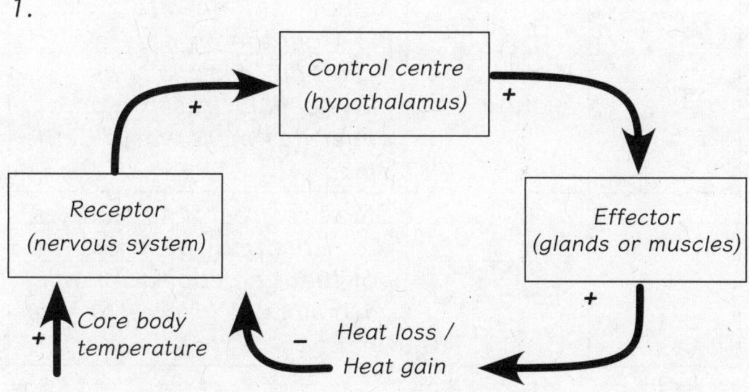

334. Control of Blood Glucose (Page 524)

1 a. The alpha cells in the pancreas are stimulated to release glucagon, which stimulates the breakdown of glycogen to glucose in the liver, which then travels to the blood to restore glucose levels.
b. The beta cells in the pancreas are stimulated to release insulin, which stimulates glycogen formation in the liver, and uptake of glucose into the tissue.

2. The organ where glucose is processed into glycogen (for storage) and the synthesis of glucose from stored glycogen.

3. One hormone acts to counteract the other: whereas insulin initiates the conversion of blood glucose into glycogen to lower blood sugar, glucagon initiates the conversion of glycogen into glucose to raise blood sugar levels.

335. Diabetes Mellitus (Page 525)

1 a. Diabetes type 1 causes high blood sugar because the cells cannot take up glucose due to difficulties with insulin production in the pancreas.
b. In type 1 diabetes, the b cells of the pancreatic islets are destroyed by the body's immune system and no insulin is produced. In type 2 diabetes, the b cells of the pancreatic islets still produce insulin but the quantities are insufficient or the cells no longer react to it.

2 a. Type 2 diabetes causes an imbalance of blood glucose, with levels often too high or too low. Common symptoms of type 2 diabetes include heart and blood vessel disease, nerve and eye damage, difficulties healing, and kidney disease. Acute low blood sugar can lead to coma, and possibly death if untreated in time.
b. Lifestyle choices: Excess intake of carbohydrate, obesity, lack of physical exercise, smoking and excessive alcohol intake.
c. A combination of blood glucose monitoring with a device that takes a small blood sample, paired with insulin injections if the blood glucose levels are too high, such as after a large meal.

336. Thermoregulation (Page 527)

1 a. Change in body temperature
b. Muscles, skin blood vessels, thyroid gland
c. Sensory nerves under the skin

2 a. Control centre
b. Stimulus
c. Effector
d. Control centre/ Effector
e. Effector

3. A combination of mechanisms: vasodilation of capillaries under the skin to release heat from the blood via the skin, release of sweat from sweat glands under the skin.

4. Newborn babies have a large skin surface to body volume, so can lose heat rapidly. As babies don't shiver like older children and adults they require another method of generating heat to maintain a suitable body temperature. The store of brown adipose tissue can be metabolized instead to generate heat.

5. Sweating releases water onto the skin surface. Water requires a large amount of heat to change from a liquid to a gas through the process of evaporation. The heat is convected from the skin surface and moves away from the body when the water is evaporated, reducing the body temperature.

6. Blood vessels, particularly the capillaries near the skin surface, can vasodilate so more blood flows through them. The heat contained in the body can be conducted to the outer surface and then lost - this will reduce overall body temperature, as cooler blood circulates back into the inner part of the body.

337. Kidneys and Osmoregulation (Page 529)

1. The microvilli in the convoluted tubules increase the absorption of substances from the filtrate, due to increased surface area.

ISBN: 978-1-99-101423-8

2 a. Filtrate is formed when the blood is forced through the glomerulus at high pressure. This process is called ultrafiltration. Filtrate originally is derived from the liquid component of the blood carrying dissolved waste substances.

b. The filtrate is modified in the convoluted tubules where substances are removed from the filtrate (or added to it) as it passes through.

3. Large volumes of blood must pass through the kidneys so that metabolic wastes being transported from the body's cells can be removed from the blood before they accumulate.

338. Kidney Function (Page 530)

1. Blood enters the kidneys via the renal artery. It is forced through the capillaries of the kidney at high pressure in the glomerulus (ultrafiltration) - a network close to the cells of the Bowman's capsule which encase them, losing plasma, which is later reabsorbed, mostly in the proximal convoluted tubule after the filtrate leaves the renal corpuscle. The blood leaves via the renal vein. It has less oxygen and nitrogenous waste (urea) and fewer toxins and ions.

2 a. The salt gradient allows water to be withdrawn from the urine (allows the urine to be concentrated).

b. Salt gradient is produced by the active and passive movement of salt from the filtrate into the extracellular fluid in the medulla. The countercurrent flow within the ascending and descending limbs of the loop of Henle multiplies the osmotic gradient between the tubular fluid and the interstitial space.

3 a. Repeated urinary tract and kidney infections, kidney stones, kidney damage.

b. A build-up of waste and foreign bodies in the urinary tract that are not flushed out; salts build up into kidney stones in their crystalline structure if urine is not dilute enough to dissolve them - these then block urinary tracts, and cause pain.

339. Control of Urine Output in the Collecting Ducts (Page 532)

1. The water moves from the tubule in the loop of Henle passively, through osmosis, as a salt gradient is built up - the concentration of water is higher in the tubule than the surrounding interstitial fluid - hence it moves from high to low concentration. In the collecting duct, the water also moves passively via osmosis - but this movement is facilitated by additional aquaporin channels that embed in the tubule cells when ADH levels are high.

2. ADH release is stimulated by low blood pressure, often due to dehydration - which retains water in the system. However, if the blood pressure rises, the ADH levels need to fall so that excess water is removed from the blood.

340. Changes to Blood Supply (Page 533)

1.

Organ	Blood flow change	Reason for change of blood flow
Brain	Remains the same	Blood flow delivering oxygen and glucose must remain steady as the demand does not change regardless of activity level.
Muscles and tissue	Increases significantly	A large demand for oxygen and energy, via glucose, is required for vigorous exercise.
Heart (muscles)	Increases significantly	The heart needs to pump faster and with stronger contractions in order to supply muscles with blood containing oxygen and glucose.
Lungs (muscles)	Increases	Muscles around the lungs that are used for active ventilation require more oxygen/energy so they can intake more air/oxygen.
Liver	Reduces	Blood is prioritized to muscles (skeletal, heart, lungs)
Kidneys	Reduces slightly	Kidneys still need to maintain the process of excretion, especially as more cellular respiration generates more waste in the blood.
Skin	Reduces slightly	Blood redirected to muscles, but if the body becomes too hot then the capillaries will vasodilate to lose extra heat energy.

341. Did You Get It? (Page 534)

1. Pollen - from meiosis to produce gametes inside the pollen in the male anther. The pollen is light, and usually sticky, so it covers the bee when it reaches inside to obtain nectar.

2. The insects gather the pollen, mostly inadvertently as they gather nectar from the flower. Many insects tend to visit the same species of flower and may fly long distances to reach another plant. They then deposit the pollen on the flower, usually brushing against the female stigma - thus ensuring cross-pollination.

3. Codominance - the heterozygote genotype also produces a distinct phenotype that is different, usually an intermediate, to dominant and recessive phenotypes.

4. Gene linkage occurs when two genes are close to each other on a chromosome - they are unlikely to be separated during the crossing over process in meiosis and therefore are inherited together. If developing a new apple variety, the two traits are coupled together - that means if one trait is bred for, the other also occurs, and vice-versa, if removing a trait, the other one is usually lost too.

ISBN: 978-1-99-101423-8

5 a. *Fertilization to form a zygote.*

b. *The ovum contains an X sex chromosome, the sperm contains either an X or a Y chromosome - at the probability of 50/50. Which chromosome the sperm contains will determine the sex. XX will be female and XY will be male - when gametes combine.*

6. *Primordial germ cells are developed by mitosis - starting in the embryo. The primary oocyte develops by meiosis in the embryo and the process is arrested until puberty. At puberty, the primary oocyte continues developing and prior to ovulation the secondary oocyte matures into an ovum (meiosis) which is completed once zygote forms.*

7. *LH and FSH levels stimulate oestrogen release, secreted by the Graafian follicle, so increase in the first half of a menstrual cycle. Oestrogen peaks in levels at around 12 days and triggers ovulation (ovum releases) which also result in a LH surge shortly after. Progesterone levels remain low to trigger uterus lining shedding, unless the ovum is fertilized, in which case they increase to maintain uterine wall.*

8. *Human chorionic gonadotropin (hCG) is released by pregnant women soon after fertilization, in rising concentrations, and is excreted in the urine.*

Theme D: Continuity and Change

Chapter 16: Ecosystems

342. Evolution and Change (Page 537)

1. *Lamarck based his theory on the assumption that changes made during an organism's lifetime were passed down and formed the basis of evolutionary change. The paradigm shift to Darwinism was due to understanding that only inherited physical traits (phenotype) could be passed on to the next generation.*

2. *The wide variety of niches in an ecosystem, with their unique abiotic and biotic factors, provides an opportunity for differently adapted species of insects. The selection pressures in each of these niches are different and this results in the 'fittest' individuals surviving to increase the frequency of their type/alleles. This causes differences in populations over time.*

3. *Through detailed observations, he could see the differences between related populations/species in different geographical areas. He could also see variation in populations of the same species but recognized their similarities and theorized that the 'sameness' of features were passed down to offspring.*

343. Variation and Natural Selection (Page 538)

1. *1) Overproduction of the population,*

   2) genetic variation in the population,

   3) competition for resources and survival of those with more favourable variations

   4) inheritance of favourable variations and proliferation of individuals with these variations.

2. *Mutations*

3. *Sexual reproduction creates new combinations of alleles, through the random process of meiosis - genetic recombination, including independent assortment and crossing-over, and through combination of gametes during fertilization. The offspring contain the mixture of alleles from both parents, but due to the different combinations of genes, no offspring are identical (except identical twins) to other offspring or to their parents. Mutations create new alleles that have not existed before, and potentially lead to new phenotypes.*

4. *Selection pressure is created when there are more individuals than resources in an ecosystem. Some individuals will have an advantage due to their genotype/phenotype, i.e. camouflage and speed. This allows them to gain a larger share of resources (food, prey) and increase their chances of survival which leads to increasing their chances of producing more offspring, and allele frequency for the phenotypes increasing in future.*

344. Abiotic Selection Pressures (Page 540)

1. *Physical (non-living) factors that determine the conditions of an environment. These include the climate and associated components that determine it, such as humidity, temperature, and altitude, and the soil/substrate type, including pH. In marine ecosystems, also salinity and pH.*

2. *Tolerance to an inherited abiotic factor is a phenotype/genotype. Those individuals that are able to tolerate a changed abiotic factor more successfully, such as lower precipitation and water availability, will have a greater chance of survival, e.g. in desert cactus, by having a phenotype such as deeper roots to access water. These genes will be passed on to the next generation in greater frequency, resulting in natural selection over time.*

345. Adaptation, Survival, and Reproduction (Page 541)

1.a *Size - better able to secure resources/mates and/or avoid predation. Smaller male cichlids must move to the lesser preferred shell environment when competition is higher.*

b. *The rock and shell environments would begin to see two distinct populations with variance in the gene pool of each. Other distinct adaptations for the shell environment, such as lighter colour or behavioural change, may occur as natural selection is now based on different selection pressures, separating populations more.*

2. *The individual elephants do not lose their tusks (unless of course killed and they are removed); rather, the tuskless phenotype is passed on from their parents i.e. whether they are born tuskless or not.*

3. *1) There is genetic / phenotype variation in a population: tusked or tuskless is genetically inherited.*

   2) There is selection pressure favouring a phenotype: intensive poaching favours tuskless elephants.

ISBN: 978-1-99-101423-8

3) The favourable phenotype can be passed down to offspring at a higher proportion.

4) There is a change in the gene pool over time: increasing proportion of tuskless genotypes/ alleles.

4. A lower population, probably through poaching, increases the proportion of tuskless elephants in the population. This period shows a reduction in numbers of elephants, likely due to uncontrolled poaching as a selection pressure, as random deaths would not likely see a change in proportion of tusks. Tuskless females were more likely to survive and therefore pass their tuskless alleles onto the next generation in greater proportions.

346. Sexual Selection (Page 543)

1. The more noticeable males in a population will get preference among the potential mates (the selection pressure) - and therefore the traits that made them noticeable, such as larger and/ or more colourful plumage, or more elaborate dance, will be passed on to the next generation in greater frequency. Physical (morphological) and behavioural traits are both genetically inherited. Over time, there will be a change in the population.
2. Sexual selection results in marked sexual dimorphism because the competitive gender (usually males) has to advertise superiority as a mate to rivals and potential suitors. The evolution of these characteristics leads to increasing divergence in appearance.
3. Sexual selection may fix exaggerated characteristics because heritable female preference for them becomes more important than any correlation they might have to fitness.

347. Modelling Selection (Page 544)

1. The presence of the larger predator acts as a selection pressure to favour duller colouration in guppies to increase survival because it helps camouflage. The results appeared to agree with the hypothesis, as less dangerous predators i.e. the removal of the selection pressure, resulted in both dull and colourful fish all eventually becoming colourful.
2. Without variation, all offspring would have the same sized spots regardless of any selection pressure. Natural selection needs to act on the variation caused by differences in alleles present. Due to a change in spots in some experiments, there must have been a variety of 'spot' alleles in original population.
3. The predator fish could more easily target guppies that stood out against the background. Therefore, guppies with large spots in coarse gravel and smaller spots on finer gravel had more fitness (less predation), and therefore they had more offspring: their alleles increased in frequency over time in respective environments.
4. Sexual selection became the selection pressure once the predator was removed. Those guppies whose spots made them stand-out from the substrate were selected by females in greater frequency, thus increasing their fitness, and subsequently the frequency of their alleles in the next generations.
5. Student answer: The specific summary response will depend upon the programme and the selection pressures chosen. However, the students should be able to observe a change in allele frequency over time in many cases. In the deer mouse population, the light-coloured deer mice should increase (and the allele for the colour) over time on light coloured rock - and vice-versa for the dark coloured deer mice.

348. Gene Pools and Populations (Page 546)

1.

A	a	AA	Aa	aa
27	19	7	13	3
58.7	41.3	30.4	56.5	13.0

2. The allele frequency of 'A' has changed from 54% to 58.7% showing an increase of the dominant allele. Subsequently, the allele frequency of the 'a' allele has dropped from 46% to 41.3%

3 a. The cays have been cleared of any lizard populations by hurricane Francis, providing isolated locations to establish populations.

b. Founder lizard populations with the longest legs at the start of the experiment still had the longest legs at the end of the experiment, even though all populations have developed shorter legs. The same occurred with short leg founder populations. This showed that the characteristics of the founder population influenced the descendants.

4 a. 3/15 x 100 = 20%

b. The founder population already had a higher than normal rate of asthma (English population) and this seems to have affected the descendants in that the rate of asthma is still much higher than in the wider English population.

349. Neo-Darwinism and Modern Synthesis of Evolution (Page 549)

1 a. Yes. Darwin's theory proposed that traits were able to be passed from parent to offspring and the parent's variant of the trait was passed onto the offspring. However, he was unaware of the molecular basis for how the trait was passed. The genetic mechanism provided the explanation.

b. New theories and hypotheses are often based on previous knowledge, research, and theories. If the new evidence supports or does not disprove what was previously hypothesized, then it builds upon the knowledge. The modern synthesis builds the molecular level mechanisms into the natural selection principles.

2. Student's own answer, but should include reference key developments and contributors instrumental in forming the modern synthesis. Mechanisms:

- Malthusian competition proposed population growth is potentially exponential but limited

ISBN: 978-1-99-101423-8

by resources.

- Variation is seen within populations
- Resource availability and variation contribute to fitness and survival.
- Mutations drive the variation of the genome resulting in genetic variation
- Mendel's law of inheritance revealed offspring inherit one copy of a gene from each parent (heritability)
- Natural selection, genetic variation, and Mendelian inheritance form the basis of the modern synthesis.
- DNA structure, embryology, DNA and protein analysis further developed the modern synthesis.

People:

Students may name many scientists but following the mechanisms provided they should include:

- Stephen Jay Gould
- Gregor Mendel
- Thomas Malthus
- James Watson and Francis Crick

350. Directional, Stabilizing, and Disruptive Evolution (Page 551)

1. Stable environments favour the most common phenotypes. Instability provides a chance that extreme phenotypes might be advantageous. Predictability reduces phenotypic variability because this is no disadvantage when the environment is largely static.

2 a. Drought favoured birds with beak sizes at two extremes of the range, since these birds could exploit the small and the large seed size.

b. Continuation of the drought could lead to further divergence in beak size.

c. The selection pressures favouring a bimodal distribution in beak size would be reduced and medium sized beaks would become relatively more common.

351. Using the Hardy–Weinberg Equation (Page 552)

1.

Recessive allele:	q	=	0.1
Dominant allele:	p	=	0.9
Recessive phenotype:	q^2	=	0.01
Homozygous dominant:	p^2	=	0.81
Heterozygous:	2pq	=	0.18

Proportion of black offspring = $2pq + p^2 \times 100\% = 99\%$

Proportion of grey offspring = $q^2 \times 100\% = 1\%$

2 a. 70%

b. 42% heterozygous; 42% of 400 = 168

c.

Recessive allele:	q	=	0.3
Dominant allele:	p	=	0.7
Recessive phenotype:	q^2	=	0.09
Homozygous dominant:	p^2	=	0.49
Heterozygous:	2pq	=	0.42

3 a. 40% dominant allele

b. 48% heterozygous; 48% of 1000 = 480.

c.

Recessive allele:	q	=	0.6
Dominant allele:	p	=	0.4
Recessive phenotype:	q^2	=	0.36
Homozygous dominant:	p^2	=	0.16
Heterozygous:	2pq	=	0.48

4 a. 32% heterozygous (carriers)

b. 80% dominant allele

c.

Recessive allele:	q	=	0.2
Dominant allele:	p	=	0.8
Recessive phenotype:	q^2	=	0.04
Homozygous dominant:	p^2	=	0.64
Heterozygous:	2pq	=	0.32

5 a. 50%

b.

Recessive allele:	q	=	0.5
Dominant allele:	p	=	0.5
Recessive phenotype:	q^2	=	0.25
Homozygous dominant:	p^2	=	0.25
Heterozygous:	2pq	=	0.5

6 a. 80%

b. 32%

c. 36%

d. 4%

e. 96%

f.

Recessive allele:	q	=	0.8
Dominant allele:	p	=	0.2
Recessive phenotype:	q^2	=	0.64
Homozygous dominant:	p^2	=	0.04
Heterozygous:	2pq	=	0.32

7. The recessive phenotype can be counted in the population from observation. This allows the frequency of the recessive allele to be known. In any Hardy-Weinberg equation two of three variables are known (1 and either p or q) thus the unknown (normally p) can be found.

Thus: 1/2500 = 0.04% frequency

q = 0.02.

ISBN: 978-1-99-101423-8

$p = 1-0.02 = 0.98$

$p2 = 0.96$

$2pq = 0.039$

$q2 = 0.0004$

8 a.

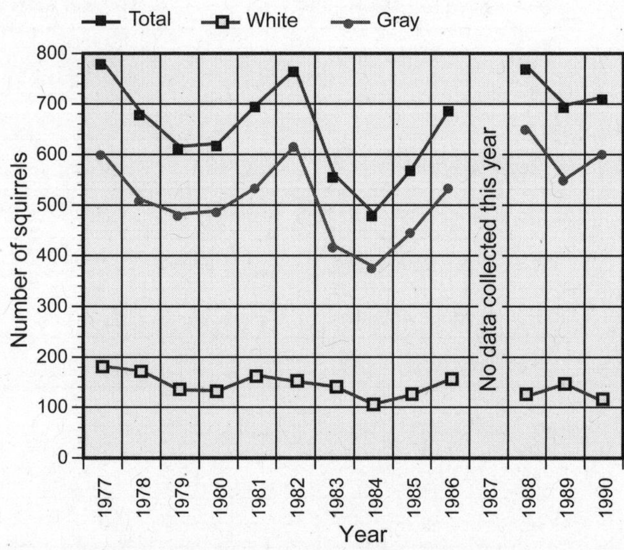

b. 784 to 484 = 61% fluctuation.

c. Total population numbers exhibit an oscillation with a period of 5-6 years (2 cycles shown). Fluctuations occur in both grey and albino populations.

9 a.

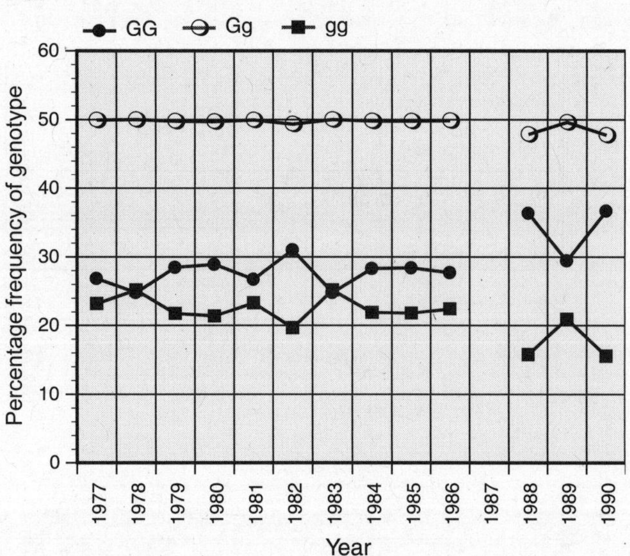

b. Homozygous dominant (GG) genotype:
Relatively constant frequency until the last 3-4 years, which show an increase.

Heterozygous (Gg) genotype:
Uniform frequency.

Homozygous recessive (gg) genotype:
Relatively constant frequency until the last 3-4 years which exhibit a decline.

10 a.

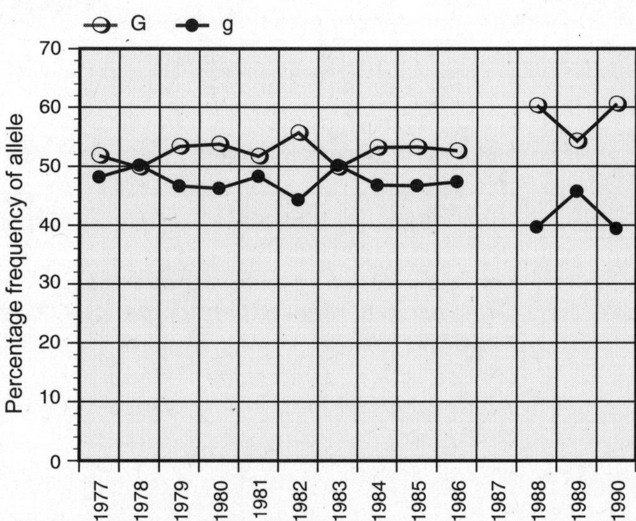

b. *Increases in the last 3-4 years.*

c. *Decreases in the last 3-4 years.*

11 a. *Frequency of alleles graph*

b. *Changes in allele frequencies in a population provide the best indication of significant evolutionary changes occurring. These cannot be deduced simply from changes in numbers or genotypes.*

12. *There are at least two possible causes (any one of):*

- *Genetic drift in a relatively small population, i.e. there are random changes in allele frequencies as a result of small population size.*
- *Selection against albinos. Albinism represents a selective disadvantage in terms of survival and reproduction (albinos are more vulnerable to predators because of greater visibility and lower fitness).*

13. *Students could explore the deer mice interactive at http://short.concord.org/lls to investigate changes in the population over time when adding or removing predators, adding mutations, and changing the habitat. Student answers will vary depending on factors selected.*

352. Artificial Selection (Page 556)

1 a. *As milk yield in Holstein cows increases, fertility decreases.*

b. *Genes for high milk yield and low fertility (or low milk yield and high fertility) are carried on the same chromosome (linked) and are very close together so that no crossing over and recombination occurs.*

2. *Milk yield will ultimately be limited by fertility (high yield cows will have lower fertility and produce fewer offspring).*

3. *The sire must carry the genes for the desirable traits, as these will be inherited by his daughters.*

4. *Selective breeding is an artificial form of natural selection usually based around the improvement of one or two traits desired by humans, in this case high milk yield. Natural selection favours traits that will increase reproductive success but this may result in a compromise*

ISBN: 978-1-99-101423-8

(best balance) between genetically linked traits, e.g. milk production and offspring production are both costly in terms of energy expenditure so in nature there will be a balance. Humans have selected for milk yield over fertility to the detriment of the latter.

5. Student answer based on discussion. Possible points of discussion include:
   - Animal welfare concerns by manipulating breeds for human ends, including health issues associated with breeding for a certain breed standard.
   - Avoiding health issues might involve legislation to maintain the gene pool while producing 'fit for purpose' dogs with good temperaments, encouraging health testing, and removing carriers of certain alleles.

6. For antibiotic resistance - the bacteria it targets is the organism that undergoes natural selection (not the antibiotic itself - which is not an organism). The selection pressure is the damaging effect of the antibiotic. Those bacteria that have phenotypes that can tolerate the antibiotic (not killed) will pass their alleles to the next generation - this is not artificial selection, as humans do not select which bacteria survive and breed.

353. Ecosystem Stability (Page 558)

1. A dynamic system is one that is constantly changing. Ecosystems, although they may appear constant, are continually changing in response to changes in the physical environment (e.g. weather, seasons) and the activities of the organisms in them.

2 a. Any suitable example such as fire, flood, seasonal drought, landslides.

b. Open pit mining, large scale forest clearance, volcanic eruptions, inundation caused by sea level rise, prolonged drought (desert formation) as caused by climate shifts.

3. Ecosystems with higher biodiversity are more stable than low diversity systems because they have a greater number of biotic interactions; these buffer them against the effects of environmental change.

4 a. The chamber with 20% algal growth media had the greatest environmental disturbance.

b. The 2% and 10% chambers

c. The ecosystem is able to resist some environmental change, but at a certain level of disturbance (20% growth medium) the system can no longer resist the disturbance and becomes unstable.

354. The Amazon Rainforest and Deforestation (Page 560)

1 a. The point reached when feedback mechanisms exceed limits where they can no longer function and the ecosystem becomes destabilized.

b. As a certain quantity of healthy rainforest is required to ensure the movement of precipitation, when a tipping point is reached, even healthy forest becomes destabilized - leading to a domino effect of rainforest death.

b. (343.0 - 313.5.4) /343.0 = 8.6%

c. The primary humid forest loss was 8.6% over 2002 to 2022 period - 25 years away at same rate.

355. Modelling Ecosystems (Page 562)

1 a. i.e warmth, light, number of snails

b. By increasing/decreasing the number of (insert factor), the mesocosm will not maintain stability (the system will undergo change).

c. Student's answer. It is expected that too much variance (by addition or removal) of any one factor will cause the system to become different (unstable) compared to the control.

2. The snails act as consumers of the plant material (detritivores), returning nutrients to the substrate.

3. Student's own answer based on their observations.

356. Keystone Species and Stability (Page 563)

1. Evidence for wolves being a keystone species can be seen by the effect they have on the elk population, and the subsequent effect this has on the vegetation in the park and the species that depend on that vegetation. The wolves have reduced the browsing elk population, which contributes to regeneration of critical plant species such as aspen, cottonwood, and willow. The elk population decline is correlated with increases in the wolf population. The increase in vegetation occurs slightly after the wolves arrive and the elk population begins to decrease (a lag for growth). Beavers also increase, which reflects their dependence on the presence of woody vegetation. The return of wolves affected not just their prey species, but all those species in the ecosystem detrimentally affected by loss of vegetation by elk overgrazing.

2 a. In the absence of sea otters, the sea urchin populations increase without constraint and they destroy the kelp forests. The kelp provides habitat for many marine species, so without the sea otters to control sea urchins the marine diversity of the area is severely reduced. Sea otters are a keystone species because their presence keeps the sea urchins under control and maintains the kelp forest on which the coastal ecosystem diversity depends.

b. The sea urchin populations would likely increase, leading to a depletion and then disappearance of the kelp forests.

357. Sustainable Harvesting (Page 565)

1. If resources are not harvested sustainably, they will not be replenished at the rate they are harvested. This will mean that, within a time period, there will not be enough individual organisms to reproduce sufficiently and the entire resource will be lost.

2. Selective logging: Advantages: reduced damage to environment, reduces crowding of trees, encourages the growth of younger trees, maintains the age distribution of the original

ISBN: 978-1-99-101423-8

forest, can be sustainable. Disadvantages: Requires time and skill to select trees for harvest. Strip cutting: Advantages: Easy to carry out. Requires little skill and planning. Cost effective. Only a small area of forest is damaged. Disadvantages: Soil in clear cut land can erode while trees are regenerating. Cleared land could be invaded by aggressive plant species (weeds), causing changes to forest composition.

3. MSY is the maximum number or tonnage of fish that can be taken without affecting the future stock biomass and replacement rate.

4 a. To determine MSY, researchers would need to know the population size, age structure, and replacement rate (population growth rate).

b. No, because they are the breeding recruits for subsequent years' biomass.

c. Ideally, fishers should use a method to catch only the smallest of the larger (mature) fish. This is not achievable with current technology, so fishers are most likely to take the largest fish and the fish population will suffer large fertility losses as a result. The fishery is difficult to fish sustainably and not viable in the long term unless managed very carefully.

358. Sustainable Agriculture (Page 567)

1. Sustainable agriculture is using farming practices that maintain crop yields while maintaining or improving ecosystem health, including water and soil quality. Farmers need to consider the total carbon inputs required to maintain their farm - reducing reliance on fossil fuels where possible - and limit use of fertilizers to areas not needed - to reduce chance of leaching into aquatic ecosystems and damage to the bacteria soil ecosystems.

2. The carbon footprint is the total net amount of carbon that a system outputs during a time period or process. In agricultural systems where the carbon footprint is positive (i.e. a net output of carbon emissions), the system becomes unsustainable in the long term as it adds to the greenhouse gases - and climate change will impact agricultural land.

3 a. Sustainable agriculture uses a variety of different crops in rotation rather than continual planting and harvesting of the same crop (monoculture). Crop rotation helps maintain biodiversity, reduces vulnerability to pests and diseases, and therefore lowers the need for pesticides.

b. Water use is a major problem all over the globe. Sustainable agriculture aims to use and store water efficiently and minimize its waste. This is achieved through protecting catchment areas (e.g. using riparian and boundary plantings), storing excess rainwater, and decreasing runoff.

c. Increasing soil health helps maintain its productivity. This is achieved by alternating nitrogen fixing crops, such as legumes, with nitrogen demanding crops such as maize. Recycling waste and manure back into the soil also boosts its organic content and productivity, and terracing and limiting tillage both help to reduce soil losses.

359. Eutrophication (Page 569)

1. Eutrophication (enrichment) is caused when nitrates and phosphates enter waterways. A consequence of the high nutrient load is excessive algal growth. Eutrophication in itself is not pollution but it is a consequence of it. The algae block light to lower depths and aquatic plants begin to die. The decomposition of the plant material, and then the algae themselves, depletes the oxygen, causing hypoxic or even anoxic conditions and causing organisms such as fish to die.

2. Oligotrophic waterways have high dissolved oxygen, high water quality (clear), low algal levels, a low level of biological production, but are likely to be stable. The opposite is true of eutrophic waterways, which are highly productive, with lower water quality measures.

3. Many land-based activities result in intentional or accidental discharges into waterways which may affect water quality. Surface runoff during rain washes chemicals, silt, and organic matter into waterways.

4. Many species in the aquatic ecosystem require a reasonable level of oxygen in the water to survive. A high BOD indicates that decomposition activity is occurring, but also depletes the water of oxygen, killing potential keystone species which results in food webs being disrupted. The high BOD also favours organisms (like algae) that can overpopulate and further damage the stability of the ecosystem.

5. The presence or absence of indicator species is correlated with certain water quality measures such as turbidity or oxygen level so they can be used to detect stream pollution. If indicators of high water quality, such as stonefly larvae, disappear from the species assemblage, this can indicate pollution (drop in water quality).

6. A BOD test could measure a series of BODs along a stretch of the water system. The point where the BODs suddenly change from low to high will indicate the entry point of the pollutant into the water systems.

360. Biomagnification (Page 571)

1. Student answer will vary depending upon the example chosen.

A suggested answer for dichloro-diphenyl-trichloroethane (DDT) biomagnification and the bald eagle is provided.

DDT is a pesticide - it is persistent and does not breakdown in the ecosystem

DDT was historically sprayed in WWII camps to prevent the spread of insect-caused disease, and was later used by farmers to prevent insects from damaging crops.

DDT washed into the waterways and taken up by aquatic zooplankton. The zooplankton were eaten by fish and the DDT became more concentrated up the food chain (zooplankton- small fish - larger fish - bald eagle). The bald eagle is the apex (top) predator and had the highest DDT levels.

ISBN: 978-1-99-101423-8

The DDT accumulated into the fatty tissue of the eagles and was not excreted. The high levels of DDT resulted in the development of eggs with very thin shells. The shell did not survive the hatching process so the birth rate of chicks decreased and the population of bald eagles fell drastically. DDT use was banned in 1972.

361. Plastics in the Ocean (Page 572)

1. The chemical bonds in plastics are not like those found in nature so there are very few organisms that can break the chemical bonds in plastic and degrade it.
2. The gyres in the oceans circulate in giant whirlpools that concentrate the debris towards their centres, therefore the floating plastic also gets concentrated. The centre of the North Pacific Ocean is in the middle of the North Pacific Gyre, so the plastic debris ends up concentrated in the water around those islands.
3. Plastic can entangle marine organisms - especially larger marine mammal and seabirds - causing drowning if caught under water, or suffocation if tied around the neck. If ingested, it could cause the animal to starve if causing a digestive system blockage - or if limiting mobility or flight to capture food.
4. Toxins attached to the surface of microplastics concentrate in consumers and cause harm. Nanoplastics can enter cells and interfere with normal metabolic processes because plastic cannot be metabolized.
5. Student's own answer.

   The answer may refer to the fact that information on social media is very accessible and is easy to share.

   They may discuss the balance and accuracy of reporting, and comment on whether the information is accurate or sensationalized. As a result, students may comment the information makes them feel overwhelmed (that the problem is too big for them to address) or they may feel the messages help them to make decisions to reduce their own plastic use.

362. Rewilding (Page 574)

1. Initially remove major pest species - provide sufficient land area for a viable ecosystem. Reduce or eliminate human interference and allow ecosystem to regenerate / rewild by itself - this may mean it goes through succession stages first.
2. Student answer will vary depending on the example chosen. The report should include detailed information on how rewilding was implemented and any impacts on the ecosystem as a result of returning the keystone species back into the ecosystem. Students may also provide information about the effect of the species removal on the ecosystem as context or background. An example answer using the reintroduction of Eurasian beavers into the UK is provided as guidance.

   Eurasian beavers became extinct in the UK in the 16th century, mainly due to human activity (hunting and habitat destruction). They were a keystone species and their disappearance had a huge impact on ecosystems.

   The dams and structures they build slow waterflow and create wetland areas which increase biodiversity, and also improve water quality as the dams act as a natural filtration system for pollutants and debris. The deeper area of water around the dam provides a water source in dry seasons. Biodiversity and water quality decreased without the Eurasian beaver's presence.

   Since 2001, beavers have been reintroduced to many areas around the UK using European stock. Trials are carefully monitored to determine the effects of reintroduction. To date, rewilding has proved successful. Waterways are again being modified by the construction of dams and from the consumption of vegetation, and biodiversity is increasing as a result. The formation of wetland also reduces flooding of farmland and human homes. Eurasian beavers obtained protected species status in 2022.

363. Primary Succession (Page 575)

1. Following glacial retreat, lava flow, landslide (exposed slip), or the eruption of new volcanic islands. Sometimes intense forest fires if they destroy even the plant seeds.

2 a. Lichens, mosses and liverworts.

b. Early colonizers erode the rock both chemically and physically (producing the beginnings of a soil) and add nutrients by decay.

3. Surtsey was an entirely new island devoid of any soil and was isolated from nearby influences such as already established vegetation or urban settlements that could accelerate the succession process.
4. Early colonizations were primarily influenced by the island's location to the south of Iceland, so the northern shores are closest to a land mass. Later colonizations were influenced by the establishment of a gull colony at the southern end of the island. The gulls would transport seeds and contribute to soil fertility.

5 a & b.

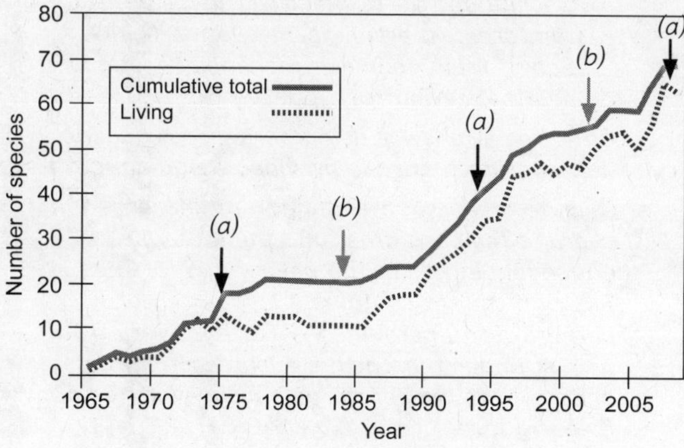

6. The gull colony caused a rapid increase in the number of plants on the island.
7. Some of the plants that make it to the island may grow one year and then die out (they may not establish or they may be out-competed during the

ISBN: 978-1-99-101423-8

succession). They are then added to the total tally but not the number of living species year to year.

364. Cyclical and Arrested Succession (Page 577)

1. *Primary succession eventually reaches a climax community that is stable and (dynamically) unchanging. Cyclic succession never reaches a final climax community - and cycles through different communities over time.*
2. *Student answer: for example the coastal ecosystem of barnacles, mussels, and algae, locust emergence, or desert wildflowers after rainfall.*

3 a. *Once drainage of the peat has occurred, completely new species occupy the new ecosystem - the previous species are likely to have become extinct (out competed).*

b. *Peatlands take a long time to develop (5000 plus years) and particular environmental abiotic conditions are required. Also, the lack of peat adapted species are likely to be completely extinct in the area - and would not repopulate the area naturally. Undoing the drainage would not occur naturally either.*

365. Climate Change: Causes and Tipping Points (Page 579)

1. *A cumulation of other connecting evidence - for example, the identification of the excess carbon dioxide in the atmosphere (above normal concentrations) can be tagged as coming from fossil fuel combustion (of which humans are the only source), and the physical processes of how greenhouse gases retain energy. Additionally, the measurements of carbon dioxide concentrations are taken from many different areas - and some measurements, like air measurement and ice core measurement, overlap.*
2. *Students' answers will vary, depending on the tipping point they have chosen. Examples could include:*

   Greenland ice sheet, Arctic summer sea-ice, alpine glaciers, coral reefs, West Antarctic sea-ice, boreal forest, El Niño Southern Oscillation, North America jetstream, Amazon rainforest, thermoline circulation (AMOC), Sahel, Indian summer monsoon, Arctic winter sea-ice, Permafrost, East Antarctic ice sheet.

   An example answer using permafrost decline in the Siberian tundra is provided as guidance.

   Climate change is resulting in an increased average temperatures, and the increase is noticeably larger at the polar regions. The Siberian tundra contains a large area of permafrost, a permanently frozen layer on or just under the Earth's surface. It persists because the temperatures hardly ever rise above freezing level.
   Frozen organic matter (and therefore carbon) lies trapped in the permafrost and is prevented from being released into the atmosphere. However, as the average temperature at the poles increases, the permafrost is beginning to melt in summer. This allows micro-organisms to become active and break down the organic matter, releasing greenhouse gases (carbon dioxide and methane). Once released into the atmosphere, the greenhouse gases contribute to further global warming and more permafrost thaws, creating a positive feedback loop. Decomposition also generates heat, resulting in further thawing. The release of greenhouse gases from a (once) locked source contributes significantly to the carbon budget and its effect on further climate change.
3. *A positive feedback cycle increases the output of an action or substance, and this output acts as a stimulus for further output. In global warming, the output is usually increased greenhouse gas emissions which cause further global warming which, in turn, through various cycles, increases greenhouse gas emissions.*
4. *The increase in one positive feedback cycle can also cause increases in another - that is, they amplify each other. For example, climate change increases the probability of drought that can cause boreal forests to die back and emit more carbon dioxide - which, in turn, causes temperature increase that affects other cycles.*
5. *Tipping points are the result of stability - and the ability of ecosystems to compensate for changes - to break down. Once some systems have gone past a tipping point, they are self-promoting in positive feedback cycles that continue, regardless of external influence, e.g. lowered emissions or temperature.*
6. *Student's answer: can be cycling to school, change to sustainable energy in the home or electric vehicles for the family, could be joining a climate march, reduction of meat / swap to vegetables/fruit/vegetarian diet, shopping for food etc. locally, reducing use of electricity in the home, promoting sustainability at school.*

366. Climate Change and Polar Regions (Page 584)

1. *Global warming (particularly so at the poles) is making the ice freeze later in autumn, melt quicker / break up quicker in spring, and less ice is formed. Less ice = less habitat, and competition increases / survival decreases for those that depend on it.*
2. *The sea ice allows the walrus to spread out and be close to where the food sources are - out in the shallow ocean. The spacing reduces competition and makes it safer for young to be left. Additionally, there is less distance to travel to find suitable food.*
3. *Reduction of walrus populations, movement north towards the pole where ice sheets remain longer. Adaptations to other food sources that are in warmer seas.*
4. *The area that the emperor penguin breeds is close to the feeding grounds of Antarctic fish (mainly) - those fish have adapted to the polar waters, and part of a food web. Simply moving the penguin will deprive it of its food source.*
5. *Student's answer*

ISBN: 978-1-99-101423-8

367. The Effects of Climate Change (Page 585)

1. *Food chains need an input of recycled nutrients to enable the producers (and consumers) to grow. The phytoplankton growth is greatly reduced if nutrient upwelling is reduced. This impacts the whole food chain, leaving little food for the larger organisms. Areas of ocean with no/limited upwelling tend to be ocean deserts.*
2. *The humpback whales would arrive at the feeding grounds (some with young) after a long period of not eating and females potentially use body reserves for energy. If encountering low food supplies, the humpback whales are at risk of starvation, and potentially the calves and adults die.*
3. *Student answer should include the following points:*

   Location of the breeding site: Antarctica - around 62 plus colonies on sea ice (possibly around 500,000 to 6000,000 adult birds)

   Link to the food supply/penguin mobility: Penguins need to be in close proximity to the rich feeding sites in the Antarctic oceans, where they feed off krill and fish. Emperor penguins over-winter in Antarctica; if they stayed on the mainland it would be too far for the adults to travel to get to the feeding sites. This is important to maintain the adults, but also to supply food to fast-growing chicks that need regular and large quantities of food so they can grow quickly and be ready to fledge and swim independently in summer.

   Period required for breeding to fledging of chicks: Egg laying from May-June. Chicks hatch 65 days later in mid-winter. They take four months to fledge (December to January in the Antarctic summer).

   Link to land-fast ice formation: The sea ice needs to remain frozen until chicks raised at the breeding colonies are fully fledged and their feathers are waterproof. They will drown if they enter the water before they are fully fledged.

   Current and projected fate of both the land-fast sea ice and the emperor penguin populations: 30% of the 62 colonies were impacted by early ice melt in the 2018-2022 period. In the Bellingshausen sea there was 100% early sea ice melt. Four of the five colonies in this region experienced 100% chick loss in 2022. Some colonies are responding by moving further inland for the next seasons - but continuing sea ice loss and early melt could result in eventual extinction of the emperor penguin.

4 a. *Ocean acidification refers to the decrease in pH of the oceans.*

b. *Since the 1850s, the pH of the ocean surface has fallen up to 0.2 pH units. The rate of has been accelerating since the mid 20th century.*

5. *The decreasing pH of the oceans is caused by the ocean absorbing increasing amounts of atmospheric CO2, forming carbonic acid and releasing H+ ions (acid).*
6. *Question numbering sequence error: No Q6*
7. *Question numbering sequence error: No Q7*
8. *They are particularly adapted to the abiotic (temperature/precipitation) of these areas - new areas to move to is limited and often the temperature change is faster than adaptation change. Animals may arrive in an area faster than food resources (plants etc.) can come to.*
9. *The areas may be patchy - if there is farmland between the forest, seeds will be unable to germinate in those areas. The birds and insects etc. that pollinate and disperse the seeds may not move at the same rate.*

368. Carbon Sequestration (Page 590)

1. *The capture and long-term storage of carbon, preventing carbon in the form of carbon dioxide from entering the atmosphere. In natural systems, carbon capture is through photosynthesis and stored mainly in the soil underneath (or as peat).*
2. *The most effective sequestration appears to be plantations of native trees. Even though carbon storage is slower by rewilding with natives initially, over the longer term it remains at a steady level. Non-native plantations have a drop in carbon each time they are harvested but have the added advantage of economic income from the wood which makes them a more viable use of privately held land.*
3. *Peat is dense with carbon. Acidic and high water conditions prevent the peat from decomposing and releasing the carbon. The peat can remain underground for very long time periods, especially as permafrost.*
4. *The peatlands form part of an ecosystem and with them is an entire ecosystem of other organisms that have evolved with it. Due to the unique features of the peatland system, these organisms are also likely to be unique and worth conserving. Restoration provides the added advantage of conservation of ecosystems.*
5. *Boreal forests tend to be in sparsely populated areas so there is space for their conservation. However, growth will be slow. In tropical areas, peat will form faster but will have to be in regions of native jungle. Temperate regions tend to have agricultural land so peatland would have to compete for space with other land use.*
6. *Forest regeneration already has some vegetation present, even if degraded, and a seed bank. These areas can grow back quicker - making sequestration quicker. Afforestation will need to have plant species (and other wildlife) brought in and be slower as the forest goes through succession (small pioneers hold less carbon).*

369. Climate Change and Phenology (Page 592)

1. *Any reasonable example, e.g. mistiming between predator/prey or food supplies, especially if one timing is based on temperature thresholds and the other is based on photoperiods, such as day/night length. It is important to understand the interdependency of organisms in a food chain.*
2. *Phenotypic plasticity enables the beetle to adjust to its environment, such as precipitation or*

ISBN: 978-1-99-101423-8

temperature changes due to climate change, e.g. by shortening the breeding cycle or the length of time in hibernation. If the range of these adjustments is wider than the range of the environmental change, then individuals will likely survive to pass on their genes. The beetle's phenotypic plasticity enables it to be able to survive and breed faster for the time required to begin to adapt genetically (generations), although it may increase competition in populations.

3. *Organisms that are unable to adapt quickly can face competition by those species that can, i.e. when another species that is better adapted to the warmer temperatures moves into the area and out-competes for the resources. Some species are closely aligned in phenology and are vulnerable if the main food resource availability (peak food mass) shifts out of synchrony to the peak food need, especially at crucial breeding times.*

4 a. *Peak food mass from the caterpillars is determined by temperature and rising temperatures can move the peak earlier. Breeding in great tits is a longer cycle initiated by day length and more 'locked in' to a particular time period than the food source. Peak food need occurs after peak food mass, limiting resources.*

b. *The ecosystem can become unstable. If there is an earlier peak food mass, this may induce the arrival of another species that can utilize it, out-competing and reducing the food resource for the great tit even more. There could also be a population explosion of caterpillars that eat more vegetation than can regrow.*

370. Climate Change and Evolution (Page 594)

1. *In areas of less snow from the increased temperature, the ability to hide from prey and secure food becomes a selection pressure. The reddish brown owls have a higher survival rate (fitness) and therefore have more offspring to increase allele frequency. If the reddish-brown allele reaches 100% frequency, then the grey-brown allele becomes 'extinct'. Even if more snow arrives, no chicks with grey-brown plumage can be born.*

2. *Less fat means less insulation and where there are temperature increases (due to climate change), this means that the gull needs to exert less effort to both maintain the fat and use energy to remain cool (within tolerance levels). The 'thinner' body type is genetically inherited so if thinner birds have more fitness, the frequency of the thin alleles in the population will increase over time - hence, evolution.*

371. Did You Get It? (Page 595)

1. *Variation in phenotypes that were advantageous to hot or cold, i.e. fur colour, fur length, ear length.*

2. *Selection pressures are factors that cause some phenotypes to have better chances of surviving and reproducing. Arctic fox (cold): thicker fur/insulation/ white colour. Fennec fox: (hot/dry) - light fur, longer ears radiate heat, smaller body size.*

3. *Global warming is directional - temperature increase leads to movement of traits in one way (mostly). Disruptive would indicate two distinct populations, and stabilizing would indicate less variation in traits.*

4. *Natural selection is the result of a selection pressure formed from an abiotic or biotic factor - whereas with artificial selection, pressure is from humans selecting which to breed. Artificial selection can result in some unintended alleles (ie. fertility) that are 'linked' becoming less frequent when other traits are selected. for.*

5. *A keystone species holds extreme importance to the stability of an ecosystem and if removed it can cause the collapse of the ecosystem. When the mammoth was removed (extinct/hunted), the tundra grass was no longer cropped and maintained. Different vegetation grew that did not allow the permafrost to be 'insulated' as much.*

6. *Permafrost will begin to melt if land/soil rises over a certain temperature. Climate change is causing permafrost melt and releasing large amounts of methane (a greenhouse gas) which increases warming and leads to more permafrost melting. This is a positive feedback loop - where the effect increases the occurrence of the initial cause.*

7. *Mammoth, as a keystone species, would 'terraform' the tundra to induce more 'insulating' grasslands to form instead of current small shrubs, and lower the ground temperature to retain more frozen permafrost. With less permafrost melting, the amount of greenhouse gas (methane) would reduce - and increase carbon sequestration.*

372. Summary Assessment (Page 596)

1. *C*
2. *B*
3. *A*
4. *C*
5. *A*
6. *A*
7. *A higher concentration of atmospheric CO_2 results in more CO_2 being dissolved into the ocean (warmer water dissolves more CO_2). The dissolved CO_2 reacts to form carbonic acid which releases H^+ ions to make the ocean more acidic. Lower pH means fewer carbonate ions are available to molluscs and coral to build shells as existing carbonate ions bond to the excess H+ It can make conditions intolerable for species like coral.*

8 a. *Any suitable use. You may wish to research one of the latest uses in human medicine e.g. curing rare genetic disease. Note: your answer must have a genetic component.*

b. *The deoxyribose sugar that makes the backbone of the DNA molecule is asymmetrical, and the carbon atoms are numbered clockwise from the oxygen atom. The 'end' that joins the phosphate group to the 3rd carbon is labelled 3', and the other end that joins the phosphate group to the 5th carbon is labelled 5'. This gives the strand directionality. The DNA polymerases add the 5' end of (new) nucleotides to the 3' end of the nucleotide strands.*

ISBN: 978-1-99-101423-8

9 a. *A measure of how successful the organism is at reproducing and passing on its genetic material to the next generation.*

b. *The earlier emergence squirrels face less intraspecific competition for food - so able to secure more food (likely to have greater fitness).*

c. *If the behaviour of early emergence is genetic, and the early emergers get more resources leading them to have a greater survival rate - as they have obtained more food and can grow quicker or have more energy for mating and breeding - then their alleles will be passed in greater frequency to the next generation.*

d. *Meiosis provides different combinations in offspring - so a variation in phenotype - it does this by independent assortment of chromosomes and crossing over during the process of meiosis - the gametes produces will then have a variety of genes/alleles from both parents i..e even the chromosome will have different combinations.*

10. *Negative feedback loops are control systems that maintain the body's internal environment at a relatively steady state. When variations from the norm are detected by the body's receptors, a response or output from the effectors that opposes the stimulus is classified as negative feedback. Low blood sugar (reduced with insulin to glycogen) stimulates glucagon release to convert glycogen into glucose, keeping steady levels in the blood.*

11 a. *Positive feedback Loops exaggerate any changes in the environment, moving the ecosystem away from a stable state by quickly amplifying changes in the environment - resulting in increased greenhouse gas emissions or increase in heat energy in the earth.*

b. *The diagram demonstrates the albedo effect. Warmer temperatures melt away snow cover faster or to a greater extent. Darker soil reflects less light energy (which could be reflected out of atmosphere) and instead absorbs more heat energy - which adds the earth's energy budget - causing further temperature increase.*

c. *Because of the long time that was required for the carbon storing soil/peat to accumulate - it is unlikely that the ecosystem will return to the same state after it has melted - and released the methane.*

ISBN: 978-1-99-101423-8

ISBN: 978-1-99-101423-8